Across the Great Divide

By
R. Wolf Baldassarro
with a foreword by Reverend Michael Copado

Across the Great Divide
R. Wolf Baldassarro
with a foreword by Reverend Michael Copado

FIRST EDITION

The following work collects original material from the author
first published monthly in *Pagan Pages Magazine* between 2009
and 2014. Some of the material herein has been updated
from its original edition for editing errors and outdated
information.

COVER: R. Wolf Baldassarro
Book design and layout: R. Wolf Baldassarro
Publisher's website:
    http://www.deepforestproductions.com
Copyeditor: Jayde Drozdowicz

For all with the courage to step out into the great unknown to expand humanity's knowledge and understanding

I tap on the door of the underworld
My breath forms a mist in the chill night air
as I call out across the Great Divide
Probing the veil between us
looking for some sign from the other side
A touch on my shoulder, a whisper in the wind,
or perhaps a glimpse of a lady in white
I am drawn to where Spirit and living collide
And then I hear a voice
Calling from across the Great Divide
I begin to wonder if there will come a day
When someone will be reading my name in stone
Will I speak up or run and hide
As they call out to me
From across the Great Divide
~ R. Wolf Baldassarro (2008)

Guest Introduction

"There are more things in heaven and earth, Horatio, than are dreamt of in your philosophy." - Hamlet (1.5.167-8)

If you've come to this because you're a fan the shows *Ghost Hunters/Lab/Adventures/Asylum-* whatever the franchise-, more than likely you will not enjoy this book; if you believe paranormal research involves breaking into cemeteries or old buildings and screaming at ghosts to try to "provoke" them into showing themselves; or huddling around your ghost box convinced that every squeal, squeak, and burst of static is a message from beyond the grave, you will, perhaps, also not like this book.

If, however, you are here because you are an open minded skeptic; interested in the history of paranormal research; its ties to the spiritualist religious movement of the 1800s; curious about Michigan/Detroit legends; want to know in practical terms how to scientifically, and respectfully, conduct an investigation from beginning to end; are interested in the science and philosophy behind parapsychology; or are looking for a pointed critique of the current world of ghost hunting in popular culture and the community as a whole; or want to know where the future in these subjects may lie, then you will find in these pages all that and more.

As a minister, I am not only fascinated by Wolf's accounts of the Spiritualist Church/Movement's contribution to modern day ghost hunting, and the psycho-spiritual aspects of these various phenomena but, more importantly, I am honestly mind-blown by his approach to the paranormal and the spiritual in a world where an understanding of quantum physics co-exists. My own work as a New Thought minister has involved moving beyond the 2,000-plus year old simplistic, often superstitious, verbiage and thought processes of my own Christian traditions (and most religious traditions as well) into a forward thinking, science-embracing celebration of the creation of the universe. Moving myself (and hopefully others) towards a new spiritual paradigm where "God" is not seen as a bearded, capricious man in the sky but simply one of the many names for that miraculous mystery which gave birth to the universe and to each of us; where we have a realization that all of creation, whether animate or inanimate, earthbound or stellar, is on a molecular level made up of the same (star) stuff. With that comes the understanding that the only thing that separates us from each other and the universe are the barriers and definitions that our minds have created. I believe that once we are aware of this atomic interconnectedness we have the power with our consciousness- as quantum physics has already proven- to manipulate these particles including those that make ourselves- or at least our consciousness (or soul)- backwards and forwards in time and across many dimensions, and that we do it all the time.

What this means is that the old religious metaphors and associations no longer work in this new quantum paradigm, and that includes our notions of what the "afterlife", such as "Heaven," "Hell", or "ghosts" for that matter, really are. If, after we "shuffle off this mortal coil", our atoms somehow merge again with the universe then what becomes of our consciousness, our soul? More importantly how, then, do we explain various paranormal phenomenon that has supposedly occurred, and sometimes been captured? Could it just be the past resonances of our consciousness and/or "energy" recorded in some place like our home, a place that had meaning to us, or where we died? Could parts of us, including our consciousness or soul, still be lingering in some lesser energetic or atomic form, sometimes barely perceptible to others? Could we actually go someplace else after we die fully (and holy?) formed and somehow, sometimes, manage to send "messages" back to our loved ones? Or possibly, as the "Many Worlds" theory of quantum physics seems to imply, and what Wolf talks about in the final chapter of the book; could these "otherworldly" things that many have experienced actually be occurrences where the walls between those many worlds, many dimensions, are thin?

I've often believed that what call "miracles" or "the supernatural" are actually just different applications of quantum and/or physical laws (like gravity) that we've just not yet been able to build sensitive enough devices to capture the "how" of them yet; and when we do, when we can understand them, they'll become day-to-day occurrences. Modern non-invasive micro surgery would have been thought to be something occult-like 2000 or even 150 years ago, but once we built a microscope that could see things smaller than the head of a needle and had a better idea of how the body worked, we could build gadgets to do things undreamt of before. To people back then, it would have been considered a divine miracle. Now these things are performed hundreds of times each day.

Perhaps one day an intrepid paranormal investigator armed with the ideas found in this book will build some sort of meter or camera that can record that exact moment when a "holy man" is able to, with their mind, manipulate the very atomic structure of water so it can be walked across; then once we understand how it was done, we will be able to do it, too.

And when we have something that can more precisely record "ghostly apparitions" or "paranormal" phenomenon on the sub atomic level or, better yet, what happens to our consciousness or soul at the moment of death, then maybe we will come to realize that the "Great Divide" isn't so great after all.

Maybe, dear reader, that investigator will be you.

Blessings,

Rev. Michael F. Copado, Unity Minister.

Author of the Blog "Towards a Grown Up God."

https://revcopado.wordpress.com/

*A*cross the Great Divide began in the fall of 2009 as an opportunity to share knowledge, understanding, and the tips and tricks of paranormal investigation with the world through publication as a regular column in *Pagan Pages Magazine*, which is published on the first day of each month.

It became a popular feature of the magazine and its many readers around the world.

This book collects those articles in a single volume for the first time and presents them in a singular context, along with some new material never before published. The original publication date subtitles each.

Here are a few comments that readers of the column have shared:

"Great article! I have been an investigator for years now, and still enjoy reading everything on the subject. I look forward to seeing more of your thoughts on the subject." - Jenny, paganpages.org

[on being interviewed] "To anyone interested in paranormal investigation and spirit photography, this is a must read piece!!" - Paul Michael Kane, professional photographer and author of *Captured: The Ruins of Eastern State Penitentiary*

"Extremely well written for us who have little or limited understanding of what some of these anomalies really are. I think Mr. Baldassarro has given all of us a greater look into what paranormal research is really about, and I personally come away from this feeling like I have learned from this. I will continue to follow this writer and look forward to the next article." - Natausha, paganpages.org

Across the Great Divide

Parapsychology Today
December 2009

*It seems appropriate to begin this anthology with the first
article published by Pagan Pages introducing myself to their
readership*

The scientific study of the paranormal has been
established for well over 100 years and although the
members of the field take experimentation and theoretical
discussion quite seriously, it remains the subject of ridicule
by many in the general scientific community and the
public.

While his formal education has been in mainstream
clinical psychology, R. Wolf Baldassarro has been a
participant in investigations and an ardent scholar of the
latest theories and data since the mid 1990s regarding the
field of psychical research.

Beginning in 2005 he returned to full-time paranormal
investigations and in January of 2009 created the
Paranormal Research & Information Society of Michigan
(PRISM)** to study, catalogue, and educate the public on
paranormal phenomena. Personal experiences and
scientific studies have provided Wolf not only with a
wealth of knowledge, but first-hand observations of
paranormal phenomena.

In recent years the practice of ghost hunting has
increased in popularity due to various reality television
shows on the subject, prompting interest in various
demographics- from teenagers just looking for a good time
and a cheap thrill to serious scientific inquiries and
experimentation. Many of today's so-called ghost hunters
consist of teens and young adults sneaking into cemeteries
and abandoned buildings at night with little respect for
local law, citizenry, or the deceased. Some are simply
worried about being taken seriously by landowners or lack

the knowledge of how to go about obtaining permission and the necessary permits.

Concerned with the lack of integrity and sincerity in many of these groups and the need for better understanding of the theories and data-gathering equipment involved in paranormal research, Wolf wrote a comprehensive reference book that separates the facts from the fiction and the do's and don'ts of a proper and professional investigation called "*A Ghost Hunter's Field Guide*" (April 2009).

The specific genre and subculture of interest is still quite new in literary markets, with many books simply recounting narratives of case studies or displaying a limited understanding of and amateur knowledge of the terms and tech. This concise field guide stands apart from so many others in the genre by providing not only a concise history of parapsychology but the wide range of terms, theories, phenomena, tools, and technology used and encountered in parapsychological study in an unbiased manner through examining both the pros and cons of popular tools and analytical techniques. Being an experienced and practicing shaman, Wolf even delves into working with spirit communication and psychic shielding.

The book also includes local legends of the Detroit area- such as the ghosts of Belle Isle- and a sample of investigations he has either been a part of or that are of particular interest to further explore. Step-by-step instructions are given in great detail for a thorough and scientific investigation of paranormal claims by someone educated and actively participating in the field of study from the founder of Michigan's premier ghost hunting and paranormal resource organization. As such, *A Ghost Hunters Field Guide* would not only be an indispensible aid to those in the field, but of interest to any fan of the paranormal or local history.

International interest in PRISM continues to grow and the group's active participation on social media continues to draw attention and requests for information and investigations by other groups, fans, local businesses, and residents.

With this rising international recognition of PRISM, a vast network of ghost hunting groups around the world, scientists and theorists in the educational community, and fans of such shows as SyFy's *Ghost Hunters* and *Ghost Hunters International*, the potential for major theoretical discussions and discoveries is immeasurable.

"The easy part was writing *A Ghost Hunter's Field Guide*", Wolf explains. "The hard part is generating a buzz within the network of ghost hunting organizations around the world."

The educational and practical applications of the material are sure to become models for paranormal study and the topics generated are clearly observed in the group's tagline, "All possibilities are seen when looking through a PRISM".

Note from R. Wolf Baldassarro: Hello and greetings to the readers of *Pagan Pages*. It is my hope that readers of *Across the Great Divide* will experience enjoyable yet thought-provoking content. Each month we will explore the mysteries of the unknown together as we discuss the world of the paranormal in entertainment and in science. There will undoubtedly be controversial topics from time to time but they will be broached and discussed fairly and equally. It is my hope that we all learn from each other and benefit from each others experiences and knowledge.

**PRISM would later be named Deep Forest Paranormal Society (DFPS). The group ceased formal operations in 2013.

Some Initial Rants

*It occurred to me during the editing process that there were some articles that didn't fit neatly into the major categories and were, esentially, more rants and commentary on some of my experiences, so I have included these here.

Stump the Ghost Guy
July 2013

I wouldn't say that I have a unique point of view among paranormal researchers, but it is a rather rare one.

Most people exist huddled at opposite extremes; that is, they either have a passionate belief in paranormal phenomena, convinced by the tiniest shreds of evidence, or they think it's all a bunch of superstition and Hollywood hogwash and keep their mind locked behind a steel door.

There are a few of us in the middle, though. We are both scientist and mystic; believer and skeptic. For as much as I am a fervent supporter of paranormal happenings, be it the adventurer in me or due to religious persuasions, I am also equally skeptic of any evidence presented to me–because the scientist in me has more questions than answers.

This double-edged sword is unleashed every time someone finds out that I am a paranormal investigator and decides to play a game of "Stump the Ghost Guy."

I'll be at a social event and someone will either already know what I do or I'll innocently give them my business card–which is a great way to make friends and network; but when I see their eyes flare upon seeing those words I get that feeling like a mouse caught in a trap: "Oh, my god! You're a ghost hunter? That's so cool! Hey, let me tell you about my: experience, aunt, grandma, etc...".

Now, don't get me wrong. I love hearing peoples' stories and I'm not trying to be pretentious, but there's a time and a place and I'm not always in business mode; however, if there aren't too many distractions or I'm not preoccupied, I'm always happy to hear a good story and share in a good discussion. The problem arises when too much is expected as a result of the conversation.

The believer will finish their story, sometimes accompanied by a grainy photo on their five-inch smartphone and anxiously await some mind-blowing revelation that confirms everything they've ever believed.

When I can't give them a definitive answer they shuffle and the conversation turns to them trying to find a way out. I know what they're thinking: "This guy doesn't know squat."

The skeptic will want me to give an answer that they have pre-written rebuttals for so they can turn up their nose and say, "Ha! I knew it. It's all lies and Harry Potter nonsense."

Both scenarios always, and I do mean always, include the storyteller asking, "Well, how do you explain it, then?"

The short of it is, I can't. Nope. I can't and I won't– especially not on the spot.

I wasn't there. I didn't see all of the gathered material, which would include all of the dismissed as well as the saved evidence. I can't verify any of the many, many environmental and human factors that could impact the event. At the end of the day it's just a story. One of many that have been shared around holiday tables and campfires for generations. Unless it's a well-researched or well-known case, then as much as the believer in me may be intrigued by the possibilities, the skeptic has to say, "Sorry, I don't know what to tell you."

I also refuse to be put in a position where the storyteller's feelings could be hurt because I seemingly dismiss their claims. Look, I'd never call anyone a liar but I wasn't there and a brief conversation and a couple of tiny digital photos doesn't give me enough material to base a knowledgeable or reputable verdict on it. Speculation is what the folks on *Ghost Adventures* do. Real science looks at all of the evidence and investigates before rendering an opinion.

To expect a detailed analysis on the spot with little or no time to properly digest everything is not only pointless but also rather rude.

The only thing I can do is keep an open mind and say, "That is cool. What have you discovered in your own research?"

It's very unusual for a case to be well-documented and referenced. This is true in academic circles, let alone outside of it. Some of the stories I've listened to aren't even the experience of the teller, but third-party recollections.

Have you ever played 'telephone' as a child? That's where someone tells you something, you have to tell it to the person next to you, and they, in turn, pass it on down the line. By the time it gets to the end it's a mess–with dates, times, and locations mixed or replaced.

I read a book about ghosts of Anchor Bay, a community off Lake St. Clair, in Michigan, which was like that- a total mess. The conjectures were off the wall at best, the photographic 'evidence' a joke, and the historical records either false or grossly misinterpreted.

While we're at it, there's been too much dependence on personal anecdotes in paranormal research these days, anyway, and it really needs to stop.

There's one that springs to mind from here in Michigan, coincidentally also in the Anchor Bay area, that I'm not going to give further credence to by repeating the name- but there are plenty of readers out there who know what I'm talking about.

I only mention it because I've heard story after story for years and there's just no feasible, credible way to verify any of it. Not a shred. It's just a cool story with a local twist. It's not science, it's story time.

Local and personal ghost stories simply exist. That's it. They don't become valid evidence simply because they're repeated or have had a few similar versions. It's never 'evidence'. They certainly don't become credible just because you've cornered a paranormal investigator with your story and he or she gives their best opinion about it, thus stamping it with the word of gospel truth.

These experiences are called 'personal' for a reason. Only you can determine just how important or how much it means to you to find the answers. Do the research. Investigate and attempt to replicate the experience.

Appreciate it for what it is- a brief moment where you peered across the Great Divide into something amazing and special.

"Hey Can You Look at This?!"
March 2012

I am contacted periodically by people and asked my opinion on a variety of paranormal "evidence." While it really sparks my interest and imagination to find a valid photo, recording, or other bit of data that can lend credence to the field, the sad truth is that these incidents are few and far between. I've fallen victim to the excitement myself while on an investigation or training exercise and 'chimped' about like a fool thinking I caught something when, in fact, it was unwarranted. Nothing is more disheartening than to present "evidence" to the scientific community only to be laughed out of the room.

There's nothing wrong with that. Quite the contrary, it sharpens your skills as an investigator and you'll know what to recognize and what to look out for the next time. What doesn't kill us makes us stronger, right?

Sometimes a person is so convinced that what they hold is worth its weight in gold that they'll try anything to get you to see their point of view with an endless array of "but…but…buts."

It breaks my heart to disappoint these folks, but it happens to me more often than I'd like to admit when I am shown print after print of dust that a person is positive is an "orb." Ah…the dreaded orb factor. My only advice is to clean your living room more often.

I've said it before, and as of late it bears repeating-99.9% of these so-called "orbs" are dust or other debris being reflected back into the camera. If it is transparent in the slightest measure, then it is dust. Period. If it is generating its own light or energy, is moving in patterns that can't be explained through the laws of chaotic motion, and is solid, then you *might* have something.

Recordings are a beast unto themselves and are really difficult for me to be conclusive on for a variety of reasons.

First of all, I was not present when these were captured so I have no way of knowing who was there, what the environmental conditions were, or many other X factors than I can even comprehend at this time. I have to take all information at the word of the presenter. That's why it takes me a long time to properly analyze recordings. I have to backup the originals (remember, you should only work with copies), run them through various filters and analytics to clean up the sound, get consensus and thoughts from other team members, and sometimes even send them out to third-party forensic labs for clarification.

I'm working on just such a case right now and will be getting the results to the client next week. I'm excited because while one recording was a bust, another did produce something substantial. Once I get the client's take, I'll be able to present it on our group's site and Facebook page.

Speaking of which, last month at a Super Bowl party I was shown an image from someone's phone of a bonfire in October, 2011. The person admitted that they used some type of filter effect on the iPhone to produce it, so if any iOS users out there can help me with what exactly they did, I'd be much appreciative. I posted the image on our Facebook page to get the public's take on it. It definitely made me raise an eyebrow, plus I thought it was a damn cool picture.

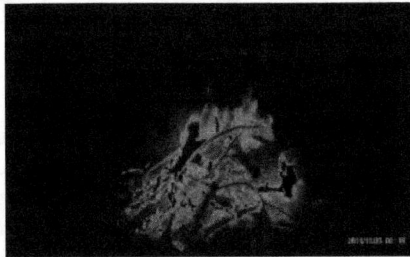

It would appear that there are humanoid figures in the flames. Call them elementals, demons, what have you- the frustrating part is that the truth may never be known.

According to the claim, it was near, or on, old Native American land. Let's face facts, anywhere you go was once Native American land, but I digress…

After asking, I was not able to get this exact photo in an unaltered state, but I was given another image from that same night and it helped shed a little insight into the situation.

So, what does this tell us? For one thing, depending on the position of the camera, any of the items in and around the campfire could produce the paranormal effects. Notice the structure in the background to the right of the image. Depending on how the flames and smoke were coming off of the fire and relative to its position, it could cause the odd-shaped humanoid figure present in the first image.

This brings up another point. I was told from the start that the first image was enhanced with what I can only speculate is some kind of spectral color filter.

Okay, take a look again at the figure on the left side of photo #1, then compare that to the smoke just barely present in the same area of photo #2. If color or heat filters where used it would force the smoke to show up in various shades of color depending on the temperature of the smoke. Voila! Instant ghost photo!

So, dear readers, what do you think? Fact or Fiction?

Paranormal Phenomenon

What kind of paranormal/ghost hunting column doesn't deal with the various phenomena encountered in this crazy world? The following section encases the many articles I have written about those subjects.

Harvests and Hauntings – Autumn in Michigan
October 2011

It's autumn again. Breathe it in. The cool air rushes in; and a patchwork of colors dot the landscape, making the world look like an open box of crayons ready to be played with. The pungent smell of dried leaves and wood fires fill the air; and our memories are pulled back to childhood images of candy corn and apples.

With the changing season also comes that carefree holiday that brings excitement and chills to child and adult alike.

Call it what you will–All Hallows Eve, Samhain, Halloween; it is nevertheless a magickal time of year when the veil between worlds is thinnest. That isn't just a philosophical point, but one of natural science; the Autumnal Equinox that ushers in the arrival of fall is marked by an equal 12 hours of day and night as the fruits of the summer harvest give way to the slumber of winter. This is the halfway point wherein we can look out across the Great Divide between the world of nature and the world of the supernatural.

Let's grab our hiking boots and gather our senses as we walk together through the bustling piles of leaves on a journey among Michigan's most haunted places.

We begin our journey in Detroit, where the General Motors plant is said to be haunted by the spirit of a man who was crushed to death in 1944; one incident recalls a worker who was saved from a similar fate by unseen forces.

Meanwhile, over at the Detroit Coca-Cola plant, a hard-lined supervisor, shot by a disgruntled worker in the 1950s, is sometimes seen or heard yelling to keep the line running when no one from management is around. So much for the mice playing when the cat's away.

Downriver from there, in Wyandotte, sits the Fifteenth Street House, where reports center on the apparition of a young girl who appears in the front window. As the story goes, there was a man who would leave for work at the same time every day, and so every day his daughter would eagerly wave to him from that window. However, one day she was not there and, thinking she just overslept, he went to leave. Upon backing his car out of the driveway, he heard a scream. In an unfortunate tragedy, she was running his lunch out to him and was struck by the car.

The Randalls of Grand Rapids met their end through a series of incidents in 1910 that culminated in a famous murder-suicide. The home immediately played host to unexplained events before being abandoned a few years later. It was eventually torn down and the Michigan Bell phone company built their office on the land in 1924. Workers would soon share tales of apparitions, noises, and doors opening and closing. The residents of Grand Rapids have had to endure decades of odd late-night phone calls that, when traced, were found to originate from inside the Michigan Bell building.

In Flint, the Cornwall family's home is now an office, but they continue to walk the halls of the building that still carries the family name. Witnesses have seen them in the old office window facing 3rd Street.

What's a story about haunted places without at least one psychiatric facility on the list? Therefore, the next stop is the Southwest Michigan Tuberculosis Sanitarium in Kalamazoo, which has benefited from a long history of stories associated with it. The abandoned hospital had tales of red lights seen filling the hallways, unexplained noises, and even writing on the walls appearing in empty rooms. Locals claim that different things happen every night including various apparitions in the windows and report hearing muffled screams and cries coming from the buildings at night. All that remains now of the sprawling $2.5-million complex is one building, with its 1895 Queen Anne water tower serving as a 175-foot tombstone for the souls who roam the grounds.

The Battenfield House in Fife Lake was the residence of one of Michigan's most well-known mass murderers. The owner loved to attend social events; and to that end she poisoned several family members, using the funerals as a means of providing the social contact she so craved. The reported paranormal goings-on include burning flames seen on a stairwell post but no burn marks or heat result from the activity.

In a little-known place located at the northeastern tip of the Upper Peninsula, a few miles north of Paradise, Michigan, is the town of Sheldrake. It is a ghost town today, figuratively and literally. It's so small you won't even find it on a map, and the few people who still reside there do not discuss the hauntings.

The town has suffered an inordinate amount of unexplained fires and boating accidents since being founded in the 1800s. The last one, in 1926, destroyed the town and, today, only a few buildings remain.

A visit here wields results before one even arrives. An old sea captain, wearing a cape and holding a pipe, allegedly appears on the dock when boats pass by. He is first seen from the lake and, as boats approach the shore, he slowly fades from the view of passengers.

The Palmer House reportedly has lights that turn on independently and shades that open in empty rooms. The Hopkins House involves a glowing apparition walking through it at night. A logger with heavy beard and overalls is sometimes seen on the furniture or in the doorways of the Smith House.

The most active building is the Biehl House, the people who owned the main manufacturing plant and most of Sheldrake. Voices are heard and many different apparitions have been sighted on the property- most notably a woman in a blue veil who has been known to walk beside visitors. Pictures will fall off walls, and faucets will turn on by themselves, but these can be easily explained in houses so old.

Every state, and nearly every town across America, has similar stories; the locations and events are as numerous as the fallen leaves that speckle the landscape. So as you return from our journey to your quiet, comfortable home town, ask around. It, too, has its own stories to share of forgotten, unseen residents.

As you or your children head out to enjoy hayrides at cider mills and take in the serene settings of the season, look behind you and in between the shedding trees. That chill going down your spine might not be a cool autumn wind, but the hint that you just might not be alone.

*An Exploration of Near-Death
and Out-of-Body Experiences
January 2011*

Let's switch gears this month away from ghosts and
the like, and take a look at another huge area of paranormal
research: the Near-Death, or Out-of-Body, Experience
(NDE/OBE).

Hollywood likes to toy with the notion of life-after-
death and the Out-of-Body Experience- sometimes in a
beautiful way, such as in 1998's *What Dreams May Come*,
and the 1980 classic *Resurrection*. I'd like to note that
Resurrection was the first movie to base its screenplay not
on superstition and cultural references, but on the research
and ideas of Raymond Moody, whose 1975 book *Life After
Life* led to the foundation of the International Association
for Near-Death Studies (IANDS) in 1981.

Stories of NDEs go back as far as recorded history,
and today, according to a Gallup poll, as many as
eight million Americans claim to have had an NDE. Some
researchers believe that number to be underestimated due
to many individuals being hesitant to talk about their
experiences for fear of ridicule and rejection.

These events are usually reported after an individual
has been pronounced clinically dead or otherwise very
close to death–thus the term Near-Death Experience. Many
NDE reports, though, originate from events that are not
life-threatening at all. With modern developments in
cardiac resuscitation techniques, the number of reported
NDEs has increased. Many in the scientific community
regard such experiences as hallucinatory, while paranormal
specialists and some mainstream scientists argue them to
be evidence of an afterlife.

The typical NDE includes a sense or awareness of
being dead; a sense of peace, well-being and painlessness;
and an observation of one's body from outside the self that
even includes detailed accounts of medical personnel
performing revival efforts. A "tunnel experience"

accompanied by a sense of moving up, or through, a passageway or staircase is the most common memory along with movement toward a powerful light. Encountering deceased loved ones and "beings of light" who take the person on a "life review" are also commonly reported.

There is a darker side to these experiences as well. A long-time reader of this column sent me a story of how her mother worked for 40 years in a nursing home. As you could quite imagine, she saw several patients leave this world over such a long period of time. "One gentleman," she said, "coded, and when they brought him back he was terrified; there was no 'light at the end of the tunnel'. Instead he described dark entities waiting for him." She noted that this man was gone for more than five minutes and postulated that what he was doing prior to his "pause" in living may have had something to do with what he saw; or perhaps his religious choice was a factor in the incident.

Such incidents remind me of the Transcendental Psychology overtones in the movie *Ghost*, wherein Patrick Swayze's character was greeted by peaceful scenes and light from above, while the antagonists of the film met their end with dark, growling shadows from below who forcefully took the soul on to the next plane of existence.

Cultural or religious factors may have a roll in these experiences that could possibly determine the format of the NDE phenomenon, but several neuropsychological and other scientific theories are being put forward to explain them as well.

In a 2006 theory developed by Richard Kinseher, the knowledge of the Sensory Autonomic System is applied in the NDE phenomenon. His theory asserts that the experience of death is an extremely strange paradox to a living organism, the shock of which triggers the NDE. According to the theory, during the NDE the individual becomes consciously aware of the brain performing a scan of the whole episodic memory, including prenatal experiences, in order to find a stored experience which is comparable to the input information of death. All these scanned and retrieved bits of information are then evaluated by the mind, as it is searching for a coping mechanism for the potentially fatal situation. Kinseher believes this is the reason why a near-death experience is so unusual, because people who experience NDEs recall memories long considered lost. His theory essentially depends upon a theory of memory in which all memories are indefinitely retained.

The theory also states that NDEs that are accompanied by out-of-body experiences are an attempt by the brain to create a mental overview of the situation and the surrounding world. The brain then transforms the input from sensory organs and stored experiences (knowledge) into a dream-like idea about oneself and the surrounding area.

Whether or not these experiences are hallucinatory, they do have a profound impact on the observer. Many psychologists not necessarily pursuing the paranormal have recognized this in recent years. These scientists are not trying to debunk the experience outright, but are instead searching for biological causes of NDEs. Their research suggested that the extreme stress caused by a life-threatening situation triggers brain states similar to REM sleep and that part of the near-death experience is a state similar to dreaming while awake. People who have experienced times when their brains behaved as if they were dreaming while awake are more likely to develop the near death experience. Stimulation of the Vagus nerve during the physical and/or psychological stress of a life-threatening situation may also trigger brain conditions where the person is in a dream-like state while awake.

Sleep researchers Timothy J. Green, Lynne Levitan, and Stephen LaBerge have noted that NDEs are similar to many of those reported during lucid dreaming. If you remember my column from back in May, 2010, lucid dreaming occurs when the individual becomes consciously aware that they are in a dream. Often these states are so realistic that they become indistinguishable from reality, even including the ability to experience textures and smells.

What's it all mean? Is it a glimpse of what's to come, or a chance to change that which is? Many people who have NDEs report an intense motivation for personal growth and change. These experiences have been known to change the attitudes, personality, and beliefs of those who have them. Psychologists see a potential for a valuable therapeutic tool in this regard due to a variety of personality shifts including a reduced fear of death, a sense of invulnerability, a feeling of importance or destiny, and a belief that they have been granted a favor. Some psychologists argue that these same qualities manifest in reckless and deviant behavior. Nevertheless, if these qualities are described by people in terms of "aliveness" as opposed to "deadness", wouldn't that make for a healthier and more meaningful life? To embrace life's possibilities instead of shutting away in a constant fear of death and low self-image would seem to be the more enlightened path.

Reminiscent of the classic Dickens' novel, *A Christmas Carol*, when a person experienced a "life review" inevitably there were profound changes from the realization of what is truly important to that individual's life. Generally, high priority values of kindness, compassion, and unconditional love for others became more prominent. Low priority values such as money, competition, and power were rejected afterward. It would seem that the story of Scrooge is an NDE account. Charles Flynn quoted a subject of his study as saying, "The superficial aspects of my life, what I had accomplished, owned, and known, were consumed and rendered unimportant. However, those acts in which I selflessly expressed love or concern for my fellow men were glorified and prudently inscribed in the record, with total disregard for however humble or fleeting those moments had been."

Another type of transformation is that people report a much greater concern for others and a communal well-being resulting in a greater willingness to accept others and to be less judgmental of others.

In my other writings, I expressed this concept once with the quote "Life and death are linked in an eternal circle. Because of this, I do not fear death. If I live life to its fullest and walk through it in beauty and pride, then death will be the same to me."

I think a world in which we thirst for and pursue knowledge and embrace life's promise, rather than cowering in fear of the inevitable, would be a far better place to exist. But that's just my opinion, I could be wrong.

Michigan Hauntings
July 2010

I've had the fortune of visiting many interesting places and to come away with some amazing personal experiences and evidence. I've included links, where appropriate, to the photos and EVPs from these locations.

In the spring of 2008, I was involved in the investigation of a house in the historic city of Mount Clemens, Michigan.

Nestled in a quiet neighborhood, under the shade of centuries-old oaks and maples, sits a home built in the 1860s. With such a long and profound history of the house and its neighborhood, it was not surprising that tales of supernatural events would surface.

The homeowner reported an astounding amount of odd occurrences in and around the home. On quiet summer nights the sound of horse-drawn carriages can be heard galloping down the street. In the house itself, there are reports of numerous interactive spirits including men, women, and children. One of the key entities is kindly referred to as "Victoria," who goes into all areas except the kitchen; another more malevolent presence is confined to an upstairs closet. The owner notes that most 'entities' are gentle and treated as extended guests.

Occupants and visitors see shadows and lights, and experience feelings of being watched or being in the presence of others; lights flicker–especially at nightfall (it should be noted that the home was completely rewired up to code in 2006); and the sound of children laughing can be heard.

While on an initial investigation of the home, I felt physically and emotionally pulled to the back of the house. Upon questioning this, I discovered that other visitors to the home had similar experiences. Objects have been thrown both in general and at people and any type of remodeling or painting stirs up activity. Technology also seems to draw activity.

During the preliminary investigation, dining room lights and stand-alone lamps did, indeed, begin flickering as the Sun went down for roughly an hour and faint whispers were heard. A presence/cold spot also quickly passed around me and into the hallway as I was given a tour.

During the formal investigation while in the former back servant's quarters a whistling sound was heard by investigators and did appear on audio recordings. Also in the servant's quarters while conducting an EVP session the chains on the ceiling fan began to move by command to specific questions and answers. A cold spot with no traceable origin was documented in the children's/guest room.

The most shocking documentation was a vortex/plasma light seen in a photograph of the kitchen as well as an EVP, during which the question was asked, "Can you tell me what year it is?" Immediately afterwards a very faint, gruff male voice says "Seventy Nine!"

There were no shortages of personal experiences, either. Another investigator, while in the children's room, saw a black shadow three separate times while looking in a mirror.

I was conducting an EVP session in the servant's closet that had the reports of a malevolent entity. Upon ending the session and preparing to leave, I felt a pull. When I turned on my flashlight, a long string of yarn originating from behind boxes and bags was wrapped around my flashlight and leading back into the pile of boxes. I didn't make any exaggerated movements during the session and was sitting in the same position the entire time.

The owner noted that most activity ceased immediately for a period of several weeks after the night of investigation but has since resumed a normal level of interaction.

Located on Kidder Road, in Bruce Township, Michigan, sits Goodrich Cemetery. This "final resting place" is home to some of the most interesting experiences, photographs, and EVPs I've experienced.

Hidden way in the back, almost forgotten, stands a lone obelisk, the Worden family marker. Perhaps coincidentally, this was also the location of plasma lights captured one night. The following pictures were taken mere seconds apart.

EVP recordings of a very deep and menacing male voice have been captured on very active nights. I've provided links to the best ones: Goodrich 08/20/2007, Goodrich Growl, "He's Mine".

While orbs are rarely evidence of spirit activity, the following picture from Goodrich is quite interesting.

Memphis Cemetery, in Memphis, Michigan, is the final resting place to some of the most prominent names in the history of Saint Clair County, with citizens interred from the founding days to the present.

There are hundreds of stories revolving around the cemetery and many more centered within the town itself- from long dead soldiers of early American wars to jilted lovers out for revenge.

The most well-known legend is of the Witches' Ball. At the very back of the cemetery, where some of the oldest headstones reside, there is a huge black marble stone that locals call the "Witch's Ball". Many children are buried here from the age of a few days to adolescence. According to locals, if you get close enough you can see faces and shadows in the stone; voices have also been heard and apparitions have been seen around the area.

One story came from older members of the community who said that after having heard noises from the ball a few too many times they went out with axes, pitch forks, and other farm tools and took turns beating up the marble orb, claiming it hasn't been active since. No record exists of this desecration, but there are chip marks, scratches, and dents on the stone that would coincide with it being struck by such tools.

From my very first visit to Memphis I've walked away with numerous EVPs of young women with clear and distinct words, some of which are direct answers to questions. These are some of my best EVPs to date: "Bridgit", "Date" or "Robin Day", girl laughing, "Why don't you believe me".

Michigan, just like any other location across the planet, has a long and complex history. It is only logical that urban legends, folktales, and other stories are told from generation to generation. As with any urban legend, there is some underlying truth to the story but time and interpretation have added to the mythos.

Belle Isle is a city park located in the Detroit River and open to the public. First off, we have the classic "honk your horn" legend. The story goes that if you drive your car onto a bridge on Belle Isle, turn your engine off, and honk your car horn three times, a spirit will appear from the woods, motioning for you to follow her. There have never been any reports of anyone following her into the woods. The ghost of Belle Isle has a couple different versions to the story: either there is a certain bridge, or any bridge, on the island will call this spirit. One version even mentioned she was an elderly woman.

In the early days of Troy, Michigan, Henry Blount purchased land in the 1820's near Long Lake (18 Mile Road to those outside of Oakland County) and Rochester Road and built a sprawling two-story home. He and his wife, Elizabeth, raised seven children in this home. Two of Blount's grandsons, Harry and Frank, continued to farm the land after his death. This is now known as the Sylvan Glen Golf Course.

In the early 1900s, the house was modified into separate living quarters for three maiden aunts of the Blount family and descendants continued to live in the home until May 13, 1924, when the land was sold to develop the golf course. It was then remodeled into a restaurant donning several names through the twentieth century including The Double Eagle, The Wooden Horse, and Shark Creek Inn. The former Blount family home has been Camp Ticonderoga since 1996.

The Ghost of Camp Ticonderoga is referred to as Hannah and the legend has been passed on through employees of the various restaurants that have occupied the Blount House. According to their story, Hannah hung herself in one of the upstairs bedrooms and now haunts the building. Strange noises have been heard and unexplainable things have been witnessed, such as doors slamming shut, lights turning on and off, and rattling dishes and objects falling for no reason. Employees will not close by themselves.

Located outside of Westland along Henry Ruff Road, and just down the street from the old grounds of the former Eloise Mental Asylum, is Butler Cemetery; also known as William Ganong Cemetery, it is widely considered the "most haunted cemetery in Michigan".

The cemetery is now neglected and overgrown with weeds and other debris. A wire fence that runs around it is grown over with vines and a rusty gate is broken at the entrance.

A witness reported encountering a woman in white crossing the road in front of the cemetery. He swerved to avoid her and she vanished right in front of his eyes. A year later, he saw the same woman again in the graveyard itself. He claimed to see her standing next to a tall monument and nearby was another apparition of a man wearing a uniform. He stopped his car for a closer look and the two figures faded away.

The stories of ghosts at Butler continue today and researchers have pointed out that there have been an inordinate number of auto accidents along that stretch of road near the cemetery, perhaps due to sightings of this lady in white.

Located on Michigan Avenue in Westland, Michigan, and coincidentally a few streets away from Butler Cemetery, stands Eloise. This was once one of the largest mental hospitals in the country. Eloise opened in 1839 as the Wayne County Poorhouse to house the mentally ill. There were many reports of patient beatings as well as patients being housed in overcrowded, unsanitary conditions. The majority of it has now been torn down or made into office buildings. The few buildings that do remain on the property of the old Asylum are the D building–now known as the Kay Beard building-, the fire house, the power plant, and the bakery. It is said that the ghosts of many of the tormented patients walk the halls of this asylum. Voices are heard in the empty halls; lights are turned on and off; and growls and moans are heard near the playground built for the use of the office workers' kids.

General information, history and video footage, and a map of the grounds can be found at http://www.talesofeloise.com.

Crisis Apparitions
May 2011

Have you ever seen a person or animal that was there one minute and gone the next? Then you were probably witness to a Lepke, a unique and interesting type of spiritual manifestation.

These sudden events have the appearance of a solid, living person and may even converse with someone; then, just as suddenly, vanish. Such apparitions are most often reported to have been encountered within, or immediately outside of, cemeteries, churches, and hospitals.

These are also commonly referred to in parapsychology as Crisis Apparitions, which we will focus on this time around.

An apparition's scientific definition is the projection or manifestation of a quasi-physical entity. There exist several different types of apparitions, though; the names and characteristics of which vary between cultures. For instance, the French call them Phantoms; here in America we call them all "ghosts."

Some other common types include Marion apparitions, which are materializations of what is believed to be the Virgin Mary; a Radiant Child is an apparition in which one sees a youthful figure glowing or surrounded by a bright aura; Shadowmen appear as a flickering black/smoky gray mass; and their misty, white counterparts are known as Vapor Apparitions.

For obvious scientific reasons, the best forms to encounter are Collective Apparitions, which are seen by more than one person. These are actually rather rare, and there are several theories surrounding their origins. The most popular theory is that if a large enough group witnesses the same event then it must be a valid example of paranormal activity, right?

Crisis Apparitions are a very interesting phenomenon because rather than being the intelligent haunting by a spirit long after they have left this world, they are seen either immediately before or after the moment of death and serve as a final farewell, or warning of what is to come, to loved ones.

This is similar to a wraith, but those are legendary ghosts that bring misfortune or death to anyone who see them.

My family has long experienced psychic events, and stories of crisis apparitions are duly noted.

My mother had related a story from her childhood when the family used to spend time down in Florida. In the middle of the night, my grandfather had a dream that his brother came to him calling his name. He awoke in a panic and decided to scribble the time on the wall. Mind you, this was back in the days when there still wasn't a phone in every house. The neighbors way down the street did have a phone and the next day said that his brother had died during the night. The time they told him was the exact time he had written on the wall.

If that didn't bring a chill to your arms, perhaps this heartfelt tale might raise an eyebrow.

One of our investigators at Deep Forest Paranormal Society gives a personal account of another variation of the Crisis Apparition phenomenon:

About 15 years ago, her son was dying in a Chicago hospital. Sitting in the room were her mother, herself, and another parent. The doctors had done all that they could do and the mood was heartbreaking, to say the least. Around 3am, all of them witnessed a very strong scent that she had recognized as her grandmother, who had passed away three months prior to this incident. When she looked to her son's bed, she saw an image of her grandmother standing on the left side of him. By 7:30, he had passed on from this world. Could she have been warning that the time had come or was she there waiting for him?

The Ghostly Side of Michigan State University
September 2011

To most folks Michigan State University represents many things: a top-rated education, a sports Mecca, even a party school. However, there's another side of the hallowed college campus that few ever see; and of those who have, most wish they hadn't.

Michigan may be the 26th state to enter the Union, but it's in the top 10 for the most haunted. Stroll with me as I explore the ghostly side of Michigan State University.

The first stop on our tour takes us to Fairchild Auditorium, which is rumored to be haunted by a young boy wandering around the stage and seats. Some report the sound of a boy laughing and the bouncing of a ball; he is often accompanied by other unidentifiable noises coming from the stage area such as loud creaks when no one is on the stage.

The stories are popular and plentiful enough that "Haunted Auditorium" fundraisers have been held in the past with tours of the building and its purported paranormal history.

Next, we have Holmes Hall, which just might have a permanent resident of unknown identity or origin on its sixth floor. Many students over the years have reported seeing a man entering the elevator, but he is never seen inside of it- or anywhere else within the building.

A reader- and former MSU student- sent in an email account of Yakely-Gilchrist Hall. In the summer of 1995, well after midnight, she awoke to the sound of someone pounding on her door. Looking out under the door she could see no one standing in front of the door, yet the pounding continued. Two security staff were called and they could hear the racket. They ran down the hall to stand in front of her door watching it rattle in its frame, with the handle ratcheting back and forth. It stopped after about five minutes.

Residents of Mason hall tell tales of the Oak Room, where a figure is often seen sitting in a chair but then gone upon second glance.

The campus green area by Beaumont Tower is known for images of couples in old-fashioned dress holding hands and walking slowly by on foggy mornings; and glimpses of a man in tails and a stovepipe hat on particularly dark nights.

Perhaps the most talked about incidents center on Mayo Hall.

Photo of Mary May from the Michigan State University Archives.

The story goes that the ghost of Mary Mayo, for whom the building is named, may wander its halls; and the building is equipped with a secret fourth-floor "Red Room" reportedly once used for Satanic rituals.

Located on the college's residential area known as West Circle, alongside other historical buildings, it is the oldest residence hall on campus and was built in 1931 as a standalone women's dormitory.

Mayo had progressive ideas about women's education that led to the first female professor of Domestic Economy and Household Science at the college. She fought for expanding the education of women and rallied for a women's dormitory on campus until her death in 1903. Rumors spread that it was murder or suicide, which have helped elevate the spooky tale to legendary status, but the truth is she died of an illness. It should be noted that she never once set foot in the building that bears her name. So what, if any, ethereal connection she has to the building remains a mystery.

No one has ever officially died in the house but there is some merit to the stories of Satanic rituals that took place on the fourth floor; these have no connection to Mayo herself and the floor has been locked for years.

Many personal anecdotes, nevertheless, pepper the internet of alumni experiences in Mayo Hall including one from a sophomore who was told stories of various apparitions walking the halls and of the lobby piano that reportedly played itself, making her sheepish about sitting at its bench.

The student who reported the incident at Yakely-Gilchrist supported these, though. She recounted a similar story of how she and a small group heard the piano playing a Bach minuet when no one was in the room.

A resident once woke up in the middle of the night and the overhead lamp in her dorm room was on. Her roommate was asleep and the girl just assumed the other had forgotten to turn it off and went back to sleep.

When they got up the next morning for class, her roommate asked if the other had awakened during the night, to which she responded in the negative.

She said that she woke up in the middle of the night and the light was on and the door was unlocked. Her roommate affirmed that she'd turned it off because she couldn't sleep with the lights on.

Whatever the truth of these reports may be, one thing is certain- Michigan State University has a rich and vibrant history. Countless individuals have walked its grounds creating more memories than there are stars in the sky. Whether in a crowded lecture hall or alone in a dark library corner, its history is shaped by each new student. Its alumni know with fondness that they are part of its history; and be it figuratively or literally, the next time a chill goes down your spine as you cram for that big exam, you just might not be as alone with your studies as you might think.

The Ghosts beneath the Mistletoe
December 2011

The days are increasingly shorter, the air chills to the bone, and nature slumbers beneath a blanket of sparkling snow. This is the time of year when we gather with friends and family to talk, share life's adventures, and relive the year's memorable moments. If you're like many folks, you're also gathered around a television to enjoy classics like *It's a Wonderful Life* and Dickens' immortal, *A Christmas Carol*.

However, take a step back and look at these holiday classics through the lens of a seasoned investigator and you'll begin to see them in an entirely new light. It is, after all, a fairly spooky ghost story wrapped around the morals of giving and sharing.

The Ghosts of Christmas Past, Present, and Future are similar to the phenomenon of "Anniversary Imprints," residual hauntings resulting from an emotional, physical, or electrical discharge that "record" an event into the atmosphere of a particular location and which usually manifest around the same time each year. Such imprints can appear non-conscious and redundant; but since the Spirits were highly interactive with Scrooge, it appears Dickens melded different aspects of the Spiritualist philosophies that were commonplace in the London of his day.

The arrival of Bob Marley on the first anniversary of his death fits the definition of a Revenant. These entities project an appearance of being distressed or misplaced; often a recently departed person who returns very briefly to make contact with loved ones to serve as an act of closure before going on to the afterlife. Perhaps the more appropriate classification for poor Marley is the Guardian, a spirit who returns to warn family members of imminent danger. These entities offer messages or aid during moments of distress to others.

The Ghost of Christmas Future is clearly a Harbinger, a ghost that brings warning of impending events.

Aside from the various spiritual entities throughout the story, some other cornerstones of psychical research play a large role in the adventure. For instance, Scrooge's journeys are what we refer to as Astral Projection, or Astral Travel. Astral Travel is the theory that a person's spiritual awareness can temporarily detach itself from the physical body, remaining connected by what is called the "silver cord," and experience things in other locations, time frames, or dimensional planes; the spiritual body and the physical body are then able to act independently of each other. That is why Scrooge travels through time and space but must return to his bedchamber to await the next spirit- and all within a single night.

This is all, of course, fiction; so what sort of real-world personal experiences provide similar events? Here are but a few anecdotes that I will share with you.

The Winter Solstice also brings with it a recurring event to residents of Lower Boscaswell (Cornwall). A lady in white holds a red rose in her mouth, then turns and walks into the fog. Some say that to see her will bring misfortune.

On Christmas Eve, in Kempston (Bedfordshire), England, local legend tells of a child that ran out of Kempston manor to greet his parents who were returning in a horse-drawn coach. He was hit by the horses and died of his injuries. Now, the anniversary of the event is marked by the reoccurring sounds of the tragic incident.

A man's mother passed away in 1964; that same year he moved from Nova Scotia to Ontario. The Christmas Eve of 1971 one string of lights on the tree, which was supposed to flash, had stopped several days before.

According to the witness, it was five minutes to midnight when the fireplace suddenly went out, the string of lights started to flash, and the other lights stopped flashing. He reported the room becoming chilled when a figure appeared in the recliner, his mother, with a smile on her face. His wife, who had never met her, reported the same thing. It never spoke; but at the stroke of midnight the fireplace lit up, the one string of lights on the tree stopped flashing, and the others started flashing again. The figure was gone and the lights on the tree never flashed again.

A woman received a call from beyond one Christmas. The phone rang and upon answering it, a familiar voice casually said, "Hello there." It was her mother's voice, who had died three years prior. The line had static noise and cut in and out.

Lewisham Station, London is the place of a crash in December 1957, caused by fog that killed ninety people and injured over one hundred. Their cries can be heard on the anniversary of the accident.

As you take in the many feasts this holiday season and enjoy the company of loved ones, take a moment to reflect on those dear departed and raise a glass in their honor–they just may be celebrating along side you and your kinfolk.

"The Ghost of Belle Isle"
April 2012

Prompted by the unseasonably warm weather, scores of people are venturing outdoors, eager to get a jump on summer fun.

For many in the Detroit area that also includes picnics and other activities on historic Belle Isle.

The park, located in the Detroit River and open to the public, is the largest island city park in the United States. In 2005, the then 101-year-old Belle Isle Aquarium was the oldest operating aquarium in the United States but was closed to cut costs in the city.

However, unknown to few besides some lifelong citizens of the Motor City, Belle Isle is much more than a serene picnic experience; as with many locations with such a long and varied past, it also has its share of urban legends and ghost stories.

The story goes that if you drive your car onto a bridge on Belle Isle, turn your engine off, and honk your car horn three times, a spirit will appear from the woods, motioning for you to follow her. I should note that there have never been any confirmed reports of anyone following her into the woods.

For generations, the tales of the Belle Isle ghost lady, also referred to as the Snake Goddess of Belle Isle, has attracted adventurous midnight riders to drive through the scary woods hoping to catch a glimpse of a woman in a long white gown. The Ghost of Belle Isle has a couple different versions to the story, too. Some say that there is a certain bridge that must be approached, while others claim that any bridge on the island will call this spirit. One version even describes her as an elderly woman.

Belle Isle has historically had a rich Native American heritage that continues to this day, so it is not surprising that most of the well-known tales involve local tribes.

Ottawa legend tells of the daughter of chief Sleeping Bear. Her beauty was so striking that he kept her hidden from the eyes of young suitors by hiding her in a covered boat on the Detroit River. One day, when he was bringing her some food, the winds–awed by her beauty–blew the covers off of her boat and it floated down the river. As it floated past the lodge of the keeper of the water gates, he also was stunned by her beauty, retrieved the boat, and brought the young beauty into his tent. This angered the winds, so the winds knocked him around until he died. Sorry for uncovering her beauty, the winds returned her to her father and begged the chief not to hide her from them again, but to let them enjoy her beauty. To protect her, and fearful that other men would follow, he placed the princess on an island in the river and sought the aid of the Great Spirit to protect his beloved daughter by surrounding the island with snakes to protect her from intruders.

There she could run free with the winds around her and the Great Spirit immortalized her by transforming her

into a white doe and letting her live out eternity on the island. When settlers learned of the island and the story, they named it Rattlesnake Island. Shortly after, it became known as Belle Isle. To this day, the maiden's spirit can be seen from time to time dancing in the wind on the island, and is often mistaken as a deer by witnesses.

As you enjoy the sunny skies and warm weather, perhaps those near southeast Michigan might find themselves spending a day on this historic island. Later, as twilight nears, those of you brave enough might stay a while in hopes of catching sight of the island's famous Lady.

Is Peche Island Cursed?
May 2012

Last month I brought to light some interesting legends surrounding Detroit's famous Belle Isle, but just off shore, a little more than a mile east, lies the small untapped wilderness known as Peche Island.

According to descendants of the French family who once settled the island for nearly 100 years, Peche Island remains untouched even while existing in the middle of urban sprawl for one very good reason: it's cursed.

The Native inhabitants tell a legend of how Peche Island was formed:

The spirit of the Sand Mountains, along the eastern coastline of Lake Michigan, had a beautiful daughter who he feared would be abducted. To protect her, he kept her floating in the lake inside a wooden box that was tethered to the shore.

The South, North, and West Winds fought over this maiden, eventually creating a huge storm, in which she drifted away to wash up at the shore of the Prophet, the Keeper of the Gates of the Lakes, at the outlet of Lake Huron. Needless to say, he was pretty happy to find the beautiful castaway.

The Winds soon found her and conspired to destroy the Prophet's lodge. The lodge, along with the maiden and the Prophet, were pulled into the water and eventually drifting through Lake Saint Clair to the Detroit River. The remnants of the lodge formed Belle Isle and the old Prophet became what is now Peche Island.

In 1789, Ontario was comprised of five regulatory districts. The Board of the Land Office for the Windsor region needed title to the island, which happened to be in the hands of the First People. A treaty was reached in 1790 for lands in the western Ontario peninsula, but it excluded Peche- possibly because the Ottawas, Chipewas, and Hurons who signed the treaty wished to retain the island as a fishing ground.

Local businessmen "failed to notice" that the island was not among the lands transferred to the Crown, and began petitioning for grants for ownership. Among them was Alexis Maisonville, who eventually obtained a de facto title to the island and it became known as Maisonville's Island.

The first permanent residents of the island were a French-Canadian family named Laforet dit Teno. Historical documents–primarily the notebook of surveyor John Wilkinson–placed their arrival somewhere between 1800 and 1812.

Direct descendant, Irvin Hansen Dit Laforet, believes they settled the island even earlier. In his article, "Peche Island: Occupancy and Change of Ownership 1780-1882" he describes how Jean Baptiste Laforest was granted the island in 1780 for his service in the British military as a guide and interpreter. No documents have ever been discovered to confirm the theory, however.

They began raising a family on the eastern shore, while sharing the island with a group of natives inhabiting the western side. According to Laforest family legend, Jean gained ownership of the island along with the exchange of livestock.

By 1834, Charles and Oliver Laforet (the use of an 's' was dropped by later generations) continued the family presence on the island. In 1857, Peche Island was officially transferred to the Crown by the Chippewas, but there were no grant applications because most locals believed that the island legally belonged to the Laforet family, as evidenced in the official minutes for the Essex County Council in June, 1868.

The last Laforets on the island were Leon (Leo) Laforet and his wife Rosalie Drouillard. Leo, the grandson of Jean Baptiste, was born on the island in 1819. He and Rosalie raised livestock, grew crops, and engaged in commercial fishing. Rosalie also made straw hats that they sold in Detroit. The couple had 12 children, the last being born in 1880.

In 1867, when a deed for the land could not be found, Leon claimed four acres when the island became part of Canada.

In 1870, Benjamin and Damase Laforet, cousins of Leon, contracted with William G. Hall, a Windsor businessman, for commercial fishing. Benjamin filed a quitclaim deed giving him squatter's rights.

Hall applied for a land patent of 106 acres in 1870, essentially seizing ownership of the entire island, except for Leo's four acres, for $2,900.

After Hall's death in 1882, his executor advertised that Hall's estate would sell the island, with fishing privileges. It was this sale that raised the question of title.

Benjamin Laforet (pictured) became involved in a lawsuit with Hiram Walker over the island.

Walker's sons purchased the property from the Hall estate on July 30, 1883, as a summer home for their father. Benjamin Laforet filed a claim on August 1 stating that he and his brother, Damase, had a one-third interest in a certain parcel of land that was described in the patent from the Crown to Hall.

The case was settled and the Hall Estate was authorized by the Supreme Court of Canada to give the Laforets a one-third share of the $7,000 that Walker's sons paid the estate.

Leo Laforet died on September 26 of that same year. According to the Laforet descendants, a group of Walker's men forced their way into Rosalie's home and made her and the oldest boys sign the deed over to the Walkers. In Laforet's article, he states that Walker's men threw $300 on the table and told Rosalie to be out by spring.

That winter, while Rosalie was in Detroit on business, someone came onto their property and ruined the winter stores. When it was time to leave, Rosalie got down on her knees and cursed the Walkers and the island. "No one will ever do anything with the island!" were her exact words, according to family lore.

Despite his sons' hopes that he would retire on the island, Hiram Walker spent years in failed attempts at commercially developing the island. He took five years to have canals dug that would allow boats to bring in supplies and to ensure the inflow of fresh water from Lake Saint Clair. Two yachts were purchased for travelling to the island from Walker's office and for cruises and parties on the river and lakes.

Walker built what was once a mansion containing some 40-54 rooms by various accounts. He planted hundreds of trees, put in an orchard, and built a greenhouse to cultivate flowers. He also created a golf course, stables, a carriage house, and installed a generator for electric lights.

It was widely thought that this "summer home", in the eyes of his sons, was actually Walker's attempt at opening a resort. His intended market, the high society of Detroit, all spent their time on nearby Belle Isle.

Willis Walker, a lawyer who had handled the purchase of the island, died soon afterwards at the very young age of 28.

In June 1895, Hiram Walker transferred the land to his daughter, Elizabeth Walker Buhl, due to his declining health. Elizabeth was no philanthropist by any means and lore tells of an incident where she denied locals from picking the island's abundant peach crop, a time-honored tradition. She had them dumped into the river, leaving people to collect them by boats.

Hiram suffered a minor stroke before dying in 1899.

Edward Chandler Walker died relatively young in 1915. Prohibition had caused embarrassment for sons and grandsons who were American, but operating a Canadian-based distillery, and not wanting to be seen as bootleggers, they sold their father's empire in 1926.

Hiram Walker & Sons distillery was purchased by Toronto's Cliff Hatch in 1926, thus ending the Walker dynasty. The Walker family leaves Walkerville and abandons the town their father founded in 1858. Some remain to this day in the Grosse Pointe area.

The ruins of Hiram Walker's mansion

Elizabeth Buhl sold the island to the Detroit & Windsor Ferry Company in 1907. The president of the company, Walter E. Campbell, stated that the island would be made into "one of the finest island summer resorts in America," and that "the big house at the upper end of the island has 40 rooms and will be easily converted into a temporary pavilion at least," according to the Nov. 11,1907 edition of the Detroit News.

Mr. Campbell apparently died in the home that same year and the property fell into ruin. In 1929, the house burned to the ground. Some Detroit residents claim that it was directly struck by lightning.

The island legally remained the property to the Belle Isle & Windsor Ferry Company, but after 1939 it transferred to the company's successor, the Bob-Lo Excursion Company. The island remained deserted except for a few picnickers, young lovers, and rumrunners during Prohibition.

It is believed that the Bob-Lo Company bought the island to deter development of competition to the Bob-Lo Island amusement park, which closed down in 1993.

Peche Island was so neglected that, in 1955, the employee who guarded the island for the Bob-Lo Company spent his spare time fishing for sturgeon, trapping muskrats, and hunting ducks without care or consequence.

Despite efforts by various local groups to have the island purchased by the government for use as a park, the Bob-Lo Company retained ownership until 1956 when it was sold to Peche Island Ltd. with plans of creating a posh residential area. With this goal in mind, the remains of the Walker house were removed in 1957. The scheme was abandoned that same year, reportedly because of a lack of suitable landfill.

Other proposals for the island followed and, in 1962, Detroit lawyer and investor, E. J. Harris, purchased it. His plan included dredging the canals and creating a ski hill and protective islands. A few years later, Sirrah Ltd. purchased the island and its water lot, despite strong resistance by many Windsor groups who wished to see the island turned into a public park. Under the direction of, Harris, Sirrah began work on an elaborate park area for the island. Several buildings, sewage, and water facilities were constructed, and phone lines were installed. The project operated for one season with ferry boats. Due to mismanagement, Sirrah declared bankruptcy in 1969, also losing the 50-acre Greyhaven estate in Detroit.

Riverside Construction purchased the island with the similar idea of developing it into a residential area or commercial recreation park that would have included a marina, but due to financial restrictions, they were forced to sell the island.

In 1971, due to lobbying by local conservationist groups, the island was purchased by Government Services with the department of Lands and Forest as the managing agency to be used by natural science students. The agency planned to spend several million dollars on the installation of nature trails, picnic shelters, and related features, but without funds the property was designated a Provincial park for administrative and budget purposes in 1974.

Currently, the island is owned by the Canadian city of Windsor as a municipal park and there are no immediate plans to develop it, apart from bathroom facilities. Other than part of the foundation of Hiram Walker's mansion, a picturesque bridge, some canals, and random piles of bricks, it looks much the way it was before the Laforets were forced off the island and Rosalie proclaimed her curse.

So, fellow explorers, did Rosalie's curse come true or not?

Most people mistakenly combine all apparitions under the blanket "ghost" moniker or laughingly think of the pop culture image of the ghost of librarian Eleanor Twitty, the very first entity the Ghostbusters were spooked by before hightailing it out of a New York library; but there are actually distinct differences in the appearance and behavior regarding these phenomena.

In proper parapsychological terms an apparition is defined as the projection or manifestation of a quasi-physical entity, the rarest being a full-body apparition.

A specific type of apparition includes the gray lady, often described in folklore as a female ghost who awaits the appearance or return of her long lost lover, child, or other person.

The Liberty Hall Mansion in Kentucky, for instance, is home to the ghost of Margaret Varick, who died of a heart attack in one of Liberty Hall's upstairs bedrooms and started appearing a few years after her death when the graves of her and several other interred family were all moved. Seen by family members, visitors, and staff, she is described as a kind, calm entity with a small frame and dressed in a gray housedress. She has been known to do some chores and her appearances never frighten or upset anyone on purpose; although, on occasion, she has annoyed when going about her business in the middle of the night, opening and slamming doors. While she has appeared suddenly in every room of the mansion, her favorite places seem to be her old bedroom and the staircase.

Haunted roads make for fascinating stories but are the most difficult to investigate due to so many contaminating factors, least of which are other cars- something that also makes these locations dangerous.

Here, at home in Macomb County, Michigan, there exists a road infamous for its gray lady.

One of the most widely talked about is Morrow Road in Algonac, Michigan. So popular is the story that a movie was filmed about the legend. Interestingly, and perhaps an indication of the validity of the legend, there are widely-differing versions of the story.

In one version, a woman walking along Morrow Road was attacked and raped sometime in the late 1800s or early 1900s. She became pregnant and left the baby by the bridge where she had been assaulted. A severe snowstorm started and she couldn't stop hearing the cries of the baby. When she went back out to retrieve the baby, it was buried in the snow and the mother died from exposure.

In another version, the child- in later years- wandered away from home one night during a snow storm. The mother ran out in search of the child, both reportedly died from exposure, and their bodies were never found. The mother now spends her afterlife in search of her lost child.

Whatever version you hear, both tell of the ghostly mother appearing to the random passerby looking for her lost child. She'll ask people, "Where is my baby?" People have claimed to have seen her, been chased in their car by her, and have heard the sounds of a baby crying. Proponents of the story claim that if you park on the road by where the bridge used to be and wait with the car off a light will appear down the road; if you speed off towards it, the light will follow and then mysteriously disappear.

These apparitions appear in all locations and at various times in countless places around the world. That man who smiled at you as you passed by on the street or

the woman walking alone at dusk–these may not be of
flesh and blood, but in fact images flickering through from
across the Great Divide desperately searching in death for
the very same things we all do in life: lost opportunities.

I'd like to tell a story that happened many, many moons ago when I was a wee little wolf cub.

When I was a child, my family went to Italy to visit relatives and were staying in a house at the top of a mountain. A relative happened to be very ill for several days at the time with high fever to the point of delusions and was passed out from sheer exhaustion and medication.

It was a quiet night with only a few of us sleeping in the house when we heard voices and saw small lights coming from a centuries-old cemetery down the way from the house. As one would expect, the adults were a bit concerned with this and wondered who was down there and what was going on but no one would even entertain the idea of venturing out there to investigate–and I wouldn't blame them!

The incident didn't last long and we inferred that it was just my father and his brothers, as they had gone outside to talk some time earlier. With many a raised eyebrow everyone went to bed.

The next morning my father was asked what they were doing in the cemetery with flashlights. He had no idea what we were talking about and informed us that he and his brothers actually went into town and so they weren't even there at the time! Let me be clear that my family and I were the only ones staying in the house, and there were no other homes in the area because it was a brand new building development.

Here's where it gets really interesting. Remember that sick relative?

He awoke in perfect health. No fever, no cough–nothing. Over breakfast, he asked who was singing last night.

"What singing?!" we all asked.

He said that he woke up at one point during the night because he heard what sounded like angels singing.

Therefore, my dear readers, what do you think happened here? A simple case of fever-induced delusion, mind over matter, spirits unsettled by the new construction in a centuries-old location, or did angels really sing a song of health and wellbeing?

In the News

The stories and incidents of parapsychological interest have often crossed the lines into the realm of journalism when events are covered by local and national media. These are some of the more interesting stories that I saw fit to discuss through the column.

"Some Debts are Hard to Pay"
April 2013

Many of us have visited a casino at least once for a number of reasons–perhaps some time in the limelight with friends at the slots amid conversation and drinks, or walking in the shadows feeding the crave of a gambling addiction. Gambling conjures up some rather strong images and opinions on all sides; this is neither the time nor place to debate those points–I only mention this because there arise, from time to time, some bizarre events that are connected with this highly-charged activity. On these rare occasions, such events can leave a chill running up your spine and a lingering wonder in the back of your perplexed mind as your psyche tries to make sense out of what just happened.

For two Metro Detroit women, a seemingly amusing time at the Motor City Casino was about to become a convergence of these possibilities as they unwittingly walked across the Great Divide into the world of the paranormal.

I am often presented with pictures and stories that set my mind reeling with questions and answers related to paranormal research–but mostly questions.

A coworker relayed a story that had made national news back in 2000 when an off-duty police officer from the Detroit suburb of Oak Park had visited the Motor City Casino and quickly ran up a $15,000 to $20,000 debt at the blackjack tables.

Perhaps from the weight of such an overwhelming loss (we'll never really know), the decorated officer stood up from the table, cried out "Noooooo!", drew his gun and put a bullet in his head while others ran for safety.

The death, believed to be the first suicide inside a U.S. gambling establishment, highlighted concerns about casino gambling. The debates consumed the nation, as tragedies are so often prone to do. Time passed. The cards kept coming and the slots kept dinging. The horror of that day was all but forgotten, at least until a few weeks ago.

My co-worker showed a news archive photo of Sergeant Solomon Bell. A happy and decorated officer–or so it would seem.

She then proceeded to pull up a photo on her smartphone, taken by her cousins in one of the restrooms of the casino. What they saw in the image would shock them to say the least:

Having been to the Motor City Casino, myself, I am familiar with how the restrooms are laid out. There is a large retaining wall that blocks and separates the entrances to the rooms- Men's to the left and Women's to the right. Once inside there is a short L-shaped entryway before actually reaching the sinks and stalls.

That being said, there is no way- in my mind- that a patron coincidentally passing by the bathrooms could possibly be reflected in the mirrors unless they were inside and standing directly behind the women pictured above.

It would seem, judging by the facial features in the image, that Sergeant Bell remains at Motor City, walking among us, and searching for a way to settle a debt that can never be repaid.

The Fine Line between Believer and Skeptic
June 2014

I've been doing a careful dance as of late on a very thin tightrope.

Lately I have been skeptical, highly critical, and utterly blatant in calling out any bullshit when it comes to ghost hunting and paranormal research news. However, I want to make it clear that I am as much a believer in the field and open-minded when it comes to the possibilities posited by the many theories as I have always been. I believe in the existence of those possibilities because I believe that we as a species do not, and can not, know everything about the universe in which we exist. Our collective understanding of how the universe works is akin to the collective understanding of science to the participants in a fifth grade science fair. Every answer only leads to more questions.

Who among you could blame me for being a bit miffed when there is a definite negative correlation between merchandising and the amount of real science these so-called reality shows share–such as a new video game based on that joke of a show, *Paranormal State*? That's right. A video game–as if science needed further mocking. Last month I talked about amateurs playing scientist, but now you can- quite literally- play scientist and think that any of the phenomena or "history" experienced in the game will translate over to the real world of science.

Legacy Interactive, A&E Television, and developer Teyon released *Paranormal State: Poison Spring* on the iPad. In the hidden-object adventure game, available to download for free on the App Store (or a full Collector's Edition priced $6.99), you can team up with the show's stars to "investigate" a "terrifying supernatural event" at Poison Spring State Park, the historical site of a horrific Civil War battle.

The gullible continue to frolic through a brightly lit open field, pointing in awe at everything; while the staunch skeptics stumble through the dark, so blinded by their narrow vision that they can't even see the light in the distance. A smart researcher knows that the road to the truth lies midway between cynicism and naiveté.

For example, do I believe in the possibility that the living can communicate with the dead in real time? Yes. However, the experiences of spirit mediums are highly subjective and the chances of proving those experiences are about as good as being hit by lightning twice.

That didn't matter to a team of ghost hunters, known collectively as Haunted Heritage, from claiming during a visit to Donington le Heath Manor House near Coalville, Leicestershire, to have made contact with King Richard III after holding a séance by the bed where he spent the night before he died. The site became a beacon for ghost hunting groups after a skeleton found in a Leicester car park last year was identified as the remains of Richard III, sparking newswires and interest worldwide.

Their night at the 700-year-old house was a complete bust (something that happens more often than not in real investigations) until they entered the chamber where Richard slept before his fall at the Battle of Bosworth in 1485.

When members of the group shouted out, "What is your name?", they say a man's voice clearly replied, "Richard."

That sounds legitimate, right? After all, he must have been the only man in the whole of written history with that name or something similar, not to mention that the mindset of the investigators had that name foremost on their minds and could have subconsciously thought or spoken the name, thus triggering the captured audio.

The medium in question, Gill Hibbert, was quick to point out that they were being careful about saying it was Richard III himself because they can't prove it and are trying to employ a historian to look into other Richards who may have lived in the residence. Of course, they define "being careful" as plastering the name and the evidence all over the internet from their Facebook page, to their website, and all over YouTube; not to mention the various news and media sites covering their story.

The group's "proof" of this contact? A response on the oft-maligned Ghost Box, a device that's been routinely slammed by numerous members of psychical research, myself included.

At least the group put their data out there for the world to mull over. Here is a link to the group's YouTube channel and the audio clip from their session. There is no mention as to whether the audio has been modified in any way (such as removing noise and hisses); had elements or volume enhanced; what the variables were; who were all those present; or what environmental and weather conditions were in effect at the time of recording.

The clip is only 34 seconds long and "Richard" only speaks at the :16 mark. I asked a credible and knowledgeable source on British history what she thought and if the speaking style would have been consistent with the late 15th century. As it turns out, whoever set this up did their research. I share in her summary that "it's bollocks, but it's well-researched bollocks."

The pattern would have been Midland or Rhotic and somewhere between Welsh and the Deleware Islands, but in her opinion the accent being from Richard III's time is believable.

From a purely scientific standpoint had the variables of the event been charged enough for his spiritual energy to become trapped in this dimensional plane, after 500 years it would have dissipated so much that the man known as Richard III no longer exists, nor could he communicate with them or anyone else.

As always, I'll leave it up to you, my informed readers, to weigh the evidence and draw your own conclusions. If you have an opinion, please share it below and join the discourse of scientific discovery.

Springtime Calls Ghost Hunters Back Outdoors
February 2011

Ridgelawn Cemetery

Well, we had a big winter thaw here in Michigan. Two feet of snow melted and the temperature even hit the 50s. The blanket of winter draws back to reveal the grass, freshly green from its long slumber. The birds return and the scents of new life are in the air.

It was short-lived, though. As I finish this article, a winter storm–complete with snow, freezing rain, and sleet descends upon the area. That brief taste of springtime, however, brings with it thoughts of sunny days and fun in the great outdoors. If you're a ghost hunter, thoughts this time of year turn to revisiting favorite cemeteries and once again traveling into the unknown and investigating buildings and places.

I love the poetic duality of cemeteries; from the serene landscapes to the ornate markers; from the sadness of a newly-dug site of a young person to the historic intrigue of a cracked and faded headstone of those long gone and forgotten to the pages of time.

I've stated time and again that if you're one of those groups that think you're serious and professional paranormal investigators, but all your troupe does is sneak into cemeteries in the dead of night to snap a few pictures and laugh and have a good time, then you're not only fools, but trespassers.

There's a big reason why laws in recent years have been established closing off these otherwise public places during the night. This was made all too clear in a recent news segment here in the Detroit area.

Vandals caused extensive damage to Detroit's historic Woodmere Cemetery this month. Rows of toppled headstones, smashed statues, and headless angels replaced an otherwise tranquil setting. One hundred and ten headstones, in all, were pushed over; some destroyed beyond repair. This is the third time vandals attacked Woodmere in the last six years. If you're caught in a cemetery at night, no matter the reason, and you have no permission to be there, then you deserve to be charged with trespassing. Enough said.

Now, I like to use cemeteries as a place to train new members; and even when I'm just out and about enjoying a nice day I'll come in with nothing more than a camera and recorder, or maybe even an EMF meter. I'll try to get a few EVPs or pictures; most of the time I'll use the weather to my advantage and snap off a library of gorgeous professional photographs; if a wandering spirit sees fit to make their presence known, all the better. I've caught enough material in cemeteries over the years to make the experience not only enjoyable on a personal level but worthwhile on a scientific one.

Some claim that paranormal activity in cemeteries is impossible. The reasons being that those interred there are long gone and any haunting will take place around the place of death, not where they were moved to a week or more after death; this being a long enough time for whatever spirit energy to cease being attached to the physical body.

Others disagree and claim that cemeteries are the most haunted spots around. Much of the photographic evidence is the subject of ridicule from serious paranormal researchers because they were often taken at night, quickly, and generally under humid conditions including mist, ground fog, and even the condensation of the photographer's own breath. The time and steps needed to rule out these environmental x-factors are simply not taken into account. Just because you're not sweating doesn't mean there isn't humidity in the air. When the temperature and dew points are within 10 to 15 points of each other formation of ground fog is highly likely.

A dirty, abandoned cemetery is going to stir up a tremendous amount of dust and dirt. Snap a flash and the resulting reflections will produce photographs that look like they were taken through a dirty car window. These are NOT the souls of the citizens of the cemetery. Nothing burns my biscuits more than being presented with picture after picture filled with these dust particles and the taker eagerly chimping away, "but look at all the spirits!" Don't waste my time, or your own, with orbs.

Aside from that little rant on orbs, cemeteries can be a great asset in many ways for researchers. You can, of course, travel freely in them during the daylight hours, but if you want to conduct nighttime investigations, you can do so legally with just a few phone calls. Contact the church, organization, or municipality that presides over the daily care and maintenance of the cemetery and seek permission. As always, be sincere and honest in your approach. If you do get the go ahead, then contact the proper authorities, and inform the local law that you will be conducting a scientific experiment in the cemetery. Get permission from caretakers IN WRITING and provide proof of that permission, along with the date, time, and a list of those group members that will be participating. They may even be willing to direct traffic around the local roads during the experiment to help reduce contamination if at all possible. It never hurts to just ask. The worst case is they simply say no. Thank them for their time and try for daylight hours. The problem here is that the increased traffic and noise levels of daytime could potentially taint any data you collect.

As with any investigation, do your research. Check local records for a history of the cemetery. Most records will, at the very least, provide a list of who is buried there. Also, look for any local events that could have caused this location to be active.

Use the daylight hours to get a map or plot from the caretaker or sketch out your own, especially if there's a particular area that interests you. You'll want to have all the landmarks and topography of the locations planned in advance for a smooth and speedy investigation.

If you or your team are thinking of doing grave rubbings, check with the caretaker first. Some very old, weather-worn headstones may be so far faded that even the light rubbing of a charcoal stick can cause further erosion. Moreover, please, do NOT do what I saw in one local cemetery. Someone had taken permanent markers or paint and lazily colored in the engraved letters on several headstones in order to make the etchings stand out. Not only is this disrespectful but also legally considered vandalism.

If it's damp, foggy, or raining, cancel and re-schedule for more favorable conditions. Any material obtained under such conditions would be inadmissible as scientific data. Besides, I sure wouldn't want to be trekking around out in the rain and mud.

As always, investigate in teams and designate a central command area and timetable. A great thing to do that will not only garner you great respect from other groups, but the gratitude and endearment of the cemetery owner is to bring trash bags to not only clean up after yourselves, but clear the landscape of other trash and debris.

Sometimes cemetery caretakers aren't interested in your data or the results, but in either case send them a professional 'Thank You' letter for allowing you to investigate. Check with them as to their wishes regarding any evidence you may have. If you obtain overwhelming evidence and make it public, it may entice those aforementioned less-respectful types to invade the cemetery, or invite vandals. No one wants another Woodmere incident on his or her hands.

So as the sunny days of spring return, keep these things in mind. Also, please, above all else respect yourselves, respect the sites, and respect the field.

Romeo Cemetery

Ghost Hunting Doesn't Involve Breaking the Law
September 2012

It seems that no matter how much I or other professional paranormal investigators lecture on the subject, some immature and amateurish yahoos out there just can't seem to get the message. If you have to sneak in under the dark of night and uninvited then you're not real ghost hunters and you deserve all that is coming to you under the letter of the law. Lately there have been more than a few stories that have crossed my desk about so-called ghost hunters having run-ins with local law enforcement.

Let's be honest. Kids have been sneaking into cemeteries for generations, but now those that are caught are using an excuse that is sadly becoming all too familiar: "But, we're ghost hunters," they exclaim.

No. You're trespassers and you're breaking the law.

It's not just cemeteries, which are usually owned and operated by a church or local historical society, that are putting up with this; privately-owned businesses and other historical locations are also facing an increase in unwanted visitors, especially after they are featured on one of the many reality ghost hunting shows that plague cable television. The owner of an abandoned mental hospital reluctantly stated that he would have to hire security guards after the stars of SyFy's *Ghost Hunters* found evidence of paranormal activity at the site. Since airing the episode, local groups and teens have been flocking to the site in hopes of a cheap thrill and capturing evidence of their own. These individuals only succeed in perpetuating the mocking of paranormal research.

Take, for example, this brief from the Bainbridge, Ohio Police Blotter:

> *SUSPICIOUS ACTIVITY, SOUTH FRANKLIN STREET: Chagrin Falls asked Bainbridge officers to assist with checking the cemetery at 3:27 a.m. on July 22 [2012]. Three Euclid women were there hunting for ghosts. No vandalism was found. The women were advised of cemetery hours and told to leave.*

I think they should have, at the very least, been gifted with citations for trespassing–which, depending on local laws, can include anything from a small fine up to, and including, jail time.

The Burnside City Council voted to deny public access to a popular park after several incidents where thrill seekers were making neighbors nervous and scared. Groups began centering on the area after a YouTube video supported evidence of a legendary haunting there.

Under the new policy, police patrols of the area would be more frequent and anyone found in the park at night would face a $5000 fine. Permits could be issued to residents who use part of the walkway to access their homes from bus stops.

Trespassing in the St. Louis-area Wildwood property, particularly in the city's parks, has become an increasing problem based on reports from the St. Louis County Police Department.

Wildwood Police Capt. Kenneth Williams said a pattern arose when residents tried to use or hide in the parks after dark; police also said some destruction of property recently occurred, prompting patrols around the park to increase after a wood carving worth $500 was stolen from the area.

This is not only scientifically unprofessional, but childish and completely unacceptable behavior from any member of society.

I'd like to note that there is a very serious problem to public safety when police have to be routinely taken off of wider patrols to focus on small areas where any mature individual with common sense and decency knows not to be. Maybe when their house is being robbed, or an accident victim's injuries might not have been so severe had an officer been closer at the time, then maybe–just maybe–they'll finally get the hint.

A new after-hours policy was established to discourage people from trespassing in parks and other areas after dark with fines for violating the ordinance being up to $1,000 or a year in jail.

Professionalism doesn't just apply to your research methods or fancy team jerseys and shiny new equipment. It extends to how you conduct yourselves on and off the field. It isn't just a love of the paranormal, but a respect for the locations and the owners of those locations.

In many cases contacting the city or church that owns the cemetery or other property and presenting your honest and objective intentions goes a long way toward garnering permission to legally access and investigate the area.

From a purely investigative nature, you could always go there during the day since we know that ghosts don't just come out at night.

I also want to point out that cemeteries, by design, are in urban areas close to well-traveled roads and residential areas. This can seriously pollute any evidence due to a large amount of X factors. Even abandoned cemeteries in secluded and neglected locations have environmental and noise pollution levels and these factors don't necessarily decrease just because it is nighttime.

Whether ghost hunting is a hobby or a serious part of your life, it should always be conducted respectfully and professionally. If you can't do that, then do public safety a favor–stay home and watch it on television.

Haunted House for Rent
June 2012

Haunted houses have been the stuff of local legend for generations; often these stories are exploited by Hollywood. Sometimes, though, there's nothing fun about the experience as the thought of sharing a home with an unexpected guest is no laughing matter.

Image © 2012 ABC News

Josue Chinchilla and Michele Callan, of Toms River, New Jersey, filed a lawsuit against their home owner, declaring that the rental property is also the residence of at least one paranormal entity and that this was never disclosed to them before moving in.

They believe they had no choice but to flee the property, along with their two children, only one week after moving in. The couple is demanding the return of their $2,250 security deposit because they claim to be witness to such activity as moving doors, flickering lights, and voices whispering, "Let it burn."

The family moved in on March 1, and immediately felt that they were not alone. By March 10, they'd had enough. That was the night that Chinchilla says blankets started inexplicably sliding off his bed and he felt an invisible grip on his arm; Callan added that she saw a "shapeless dark apparition" nearby.

They at first laughed off the activity as a trick of the senses, then tried to ignore the recurring events but, according to a report by the Asbury Park Press, when the entity began to pulling sheets off the bed and a dark apparition appeared in the bedroom they knew they had to leave.

The family checked into a hotel on March 13, where they have been staying ever since.

The home's owner refutes the entire "haunted house" lawsuit is a ruse, and counters that they are making up the story because they can't afford the $1500 monthly rent; he defends that he has rented the home for the last ten years without any issues.

In a report by the Huffington Post, the Shore Paranormal Research Society, based in the same town, investigated and later classified the activity as "paranormal," but that it did not indicate a haunting, according to the group's website. They have released a video showing a bowling pin reportedly falling over of its own accord.

Marianne Brigando, an investigator from New Jersey Paranormal Investigators, says that the home is indeed haunted and states that it is "the site of an active or intelligent haunting." She based her conclusion on the results of the oft-maligned "flashlight test", which involves communicating with an otherworldly presence via a flashlight. When Brigando asked questions, she claims an unseen force would turn the flashlight on and off- two flashes for "yes," one for "no."

To add to the controversy, the couple's pastor, Terence Sullivan, reported that he is certain a "demonic possession" has visited the house, raising the possibility that Chinchilla and Callan may have brought the ghost with them.

Alternatively, it is possible that the house does have a history, which the owner is refusing to disclose to tenants.

Consumer interest laws have arisen in recent years demanding that landowners disclose the history of a property including any reported paranormal activity, or cases of murder and suicide. If the property owner knew about the haunting, he had a duty to disclose the information before the lease was signed.

We've often seen the flip side of this, with restaurants, hotels, and other businesses, reaping the press and rewards of having 'big name' groups like The Atlantic Paranormal Society (TAPS) claim on air that a place is haunted and then sit back and watch their customer base boom.

It's arguably a material fact, but if they brought the ghost with them, then it's a haunting of another level, and the non-disclosure lawsuit would most certainly fail.

A hearing was scheduled for the end May, and both parties agreed to have their case heard on the nationally syndicated television program, "The People's Court."

Kerstin Augur, publicist for the show, confirmed that the case was to be heard before Judge Marilyn Milian in the program's New York studio.

While a studio audience was present for the taping of the program, the television courtroom is closed to the general public as well as to the media. The episode aired in September 2012.

Packing Up and Moving a Haunted House
March 2013

A rigging crew in Monroe, Iowa, spent February 19, 2013 slowly moving a 100-ton house; and that has local ghost hunters anxiously waiting to see if any paranormal activity follows the reportedly-haunted dwelling after it moves.

The home, which had to be moved after one person bought the land and another person bought the house, will be relocated about five miles from where it had existed since 1865. A photo gallery highlighting the move can be viewed on station KCCI 8's website.

Known as Maple Grove Hill, it is the oldest wood frame historical house in Jasper County and believed to still be the official residence of the original owner, Regina Long. For more than a century, her spirit has been credited with scaring off people and constantly moving items in the home.

The question on everyone's mind is if she will continue to haunt the house in its new location; the incident is also reigniting a debate in parapsychology about whether it is the location or the objects contained within it that are associated with psi phenomena.

The reason that this has been the subject of contention is due to differing opinions within the field regarding the nature of hauntings. Hauntings, as we know, are specific locations where paranormal activity occurs regularly, especially for more than a year. This phenomenon can be further broken down into the sub categories of apparition (intelligent) and imprints (non-intelligent).

Apparitions are the conscious, disembodied entities that remain after physical death that continue to walk between dimensions and are responsible for many of the incidents reported to occur in places like Maple Grove Hill.

Residual hauntings, or imprints, begin with a Flash Point—the initial emotional, physical, or electrical discharge that "records" an event into the atmosphere of a particular location. Theorized to be a psychic event that is recorded in both space and time, it continues to loop repeatedly in a particular location. Anniversary Imprints, for example, can appear non-conscious and redundant, usually manifesting within a predictable measure of time following the flash point. The intensity of this energy or recording degrades over time until the event ceases to reappear due to dissipation of the energy related to the event. This view portrays the repeating events of a residual haunting as mindless, soulless traces of past lives, not as the active, intelligent movements of spirit energies.

Sometimes paranormal activity is specifically linked to the land itself, such as Civil War soldiers that can be seen at places like Gettysburg. These sightings have been known to exhibit characteristics of both apparitions and imprints.

Other times the activity seems to be centered on the objects most commonly associated with a person and there has been evidence of activity following these objects, such as when they are moved from the original home or location to historical societies and museums.

Maple Grove Hill is to be set on its new foundation in the spring. In the meantime, a perfect opportunity exists for studying both sides of the debate and adding to scientific exploration and discovery—so it would be to everyone's advantage not to wait.

The first step would be for researchers and historians to pool the available data on reported incidents, as well as a detailed analysis of the geological properties of both the old and new locations to provide a baseline.

During this time, they should also maintain a log of any possible activity at its holding site and then follow up with a thorough investigation upon completion of the move, followed by additional evaluations at 6 months and one year after construction. After a careful analysis of all the resulting information, a clearer picture should be available to make a case for either side of the debate.

Therefore, I think there are really two questions to be answered here. First, will the spirit of Regina Long accompany the new owners and, second, will sensationalism and wasted opportunity prevail over rational and sound research?

We're watching, Iowa.

Who Left the Gate Open? The Idiots Got Out Again.
November 2013

For all of the many topics and issues within the realm of parapsychology there is one phenomenon that–pardon the pun–continues to haunt the field; a phenomenon that is predator rather than prey; and despite the most valiant efforts by some, a phenomenon that could be easily kept from causing mayhem–if it weren't for someone constantly leaving the gate open and letting it out.

I am talking about incompetence. What should come as common sense and an axiom is apparently seen by some as a hindrance; a roadblock to their fame and fortune. Moreover, unlike many topics that the origins of which are relatively unknown, there are clear methods of manifestation for this problem.

Take, for example, a recent story out of Montana about a city employee who got into trouble after she let a local ghost-hunting group set up an infrared camera in the Butte-Silver Bow County Health Department because she thinks that the office is haunted and contacted a group to catch the unwelcome spirits in action.

John DeMuary, co-founder of the Butte Paranormal Investigative Team, told the New York Daily News that he complied with the request because he is also convinced that it's haunted based on the woman's claim that "she [reported] a lot of strange things were happening and that she heard strange noises coming from certain parts of the building."

In the interest of fairness and journalistic integrity, I should point out that DeMuary is a rookie in the field and began ghost hunting just over two years ago. He has a lot to learn; and now that inexperience and arrogance has landed someone in a very real world of trouble. I can't say that I feel sorry for her in the least. Despite what she may personally believe, her actions, and the actions of the "investigative team", were completely inappropriate and unprofessional.

DeMuary did "a bit" [sic] of research and found that
the office building was constructed in the 1970s and before
that a woman had spent 80 years of her life in a house that
previously stood on the grounds. He was unable to verify
whether she died at the location but nonetheless speculated
that, "Maybe her spirit wasn't able to move on."

So on one night last August, the group snuck into the
building with the help of the employee. Their investigation
backfired when another employee turned the camera over
to police, fearing that someone was using it to spy on the
government workers.

The Butte police found no ghosts on the camera's SD
card, but plenty of normal office interactions; and the
office managers were in no way amused or sympathetic to
the situation, feeling that the incident was a violation of the
public trust. The employee who contacted the group was
given a formal written warning and another employee who
had knowledge of the situation was given a verbal warning.
No criminal charges have been filed against anyone, but I
would think it only fitting that some form of trespassing
fine be imposed on the team.

The first obvious issue to be addressed is that all
involved were not only trespassing, they were willfully
trespassing on government property. That takes enough
testicular fortitude for this world and the next. How hard is
it to understand that whether it's a cemetery, someone's
home, or a professional building, it is NEVER okay to just
walk in whenever you feel like it without the landowner's
permission?

It doesn't matter if it's an abandoned building either.
In cases of an abandoned property, sometimes contacting
the city or church that owns the cemetery or building and
presenting your honest and objective intentions goes a long
way toward garnering permission.

You should also have a client contract that explains
what each party's legal and financial responsibilities are.
Often having an explanation of what is publicly planned
for the data collected or a clause that releases the

building's owner of responsibility due to injury puts their mind at ease.

It is also relevant to mention that any seasoned and professional paranormal research group will require all members to wear photo identification while investigating or representing the group in public, even when just doing research in a library or records office. Not only does this present a more professional image but it helps clients, law enforcement, and others know who is and is not part of the group. Law enforcement has the right to request identification; and trespassing on private property can lead to fines, imprisonment, or worse. I've personally known of ghost hunting groups getting shot at when trespassing.

Furthermore, many professional buildings are, by design, in urban areas close to well-traveled roads and occupied by more than one company with their own hours of operation. This can seriously pollute any evidence due to a large amount of factors involved. Even abandoned cemeteries in secluded and neglected locations have environmental conditions and noise variables to account for that could skew results.

DeMuary said his merry little band of trespassers noticed lights flickering and thought it was "weird" and took one picture of an "orb". He also based his conclusion that the building was haunted on noises he heard from his Ovilus X, the "ghost box" that I will explore in detail later, saying that he "understood words" but "didn't know what they were saying." He just guessed that they were from a past employee of the office who died of breast cancer and who may have been trying to communicate. The choice of instrumentation, the methods of research, and the analysis techniques of this group are laughable. There's really nothing positive to say about it at all.

Orbs? Really? For as much of a joke as the TV reality ghost groups have become, even they shrug off orbs as rarely legitimate. Yet, the sad fact remains that so many amateur groups out there still try to pass off any bit of anomalous dust in the air as spirit manifestation.

I've given way too much attention to the pathetic toy that is the Ovilus than it's worth. The plastic casing of it has more value than its data.

Let this be a lesson to everyone. The more structured and professional you are in your methods, the more professional you will come off when investigating, and the more serious your data will be taken.

However, none of this seems to be of concern to the amateurs pulling the reigns of the Butte Paranormal Investigative Team. To them it's just saddle up and lock and load. It's time someone put a lock on the gate before more idiots get out.

The Only Thing Worse than Amateur Scientists Are Drunken Amateur Scientists
April 2014

Photo: *San José Library Digital Collections*

Just when I thought that the bar of integrity in ghost hunting couldn't get any lower someone went and removed it entirely.

I only allow members to bring a small water bottle on investigations, but the owners of one of the most infamously haunted locations in the country will now allow you to drink alcohol anywhere and everywhere while you play scientist and fall victim to their new amusement park.

Built in the 1880s, the Winchester House has never been registered for lodging let alone licensed for alcohol, but the current owners of the 160-room California mansion, Winchester Investments LLC, will allow guests to stay the night and drink anywhere on the 6-acre maze of false doors and stairs that lead nowhere. Sarah Winchester reportedly built the famed additions in order to confuse the evil spirits she believed were the tormented souls of those killed by the family's firearms business haunting the home.

A special use permit, which was approved by the San José planning department on March 5, 2014, will allow overnight guests to stay at the landmark site. It was made clear that their target market won't be road-weary travelers on family vacations or traditional hotel renters, but those who want the "ultimate Winchester House experience". In other words, a niche of clients predisposed to fall for their smoke and mirror act.

A reporter for the Silicon Valley Business Journal attempted to contact the owners to ask for plans that are more detailed but, conveniently, his request was ignored.

A visit to the Winchester House website reveals that no room rates have been added yet, but they have plenty of theme tours playing up the haunted history that seem better fit for Cedar Point's Halloweekends, like their "Friday the 13th Flashlight Tour". Tour rates range from $26-$65; and you can't visit any page on their website without a highly-intrusive popup ad for the books and movie that have been made about the House making it perfectly clear that they don't care whether or not the claims are real and substantiated–they're going to use it and make a lot of money.

The city also approved the remodeling of the Winchester's existing café into a full service restaurant, which will be open to the public.

Basically, this is nothing less than a clever marketing and moneymaking scheme by Winchester Investments to cash in on the legendary status of the residence by exaggerating and accentuating activity while bilking money from those foolish and gullible enough to fall for it. Well, we all know what they say about a fool and his money.

So an iconic and historical landmark will degrade into a booze-filled joyride of misfits and fools looking to play ghost hunter. I wonder how long it will be before the grounds are littered with beer bottles and a once-majestic residence falls prey to countless grubby feet and hands eroding every inch. The addition of alcohol to the equation is just asking for a lot of trouble in terms of increased police patrols, vandalism, and disruptive behavior.

I have nothing wrong with an historical site using public funds via tours to help offset the costs of maintaining beloved locations, but this goes far and beyond. It enables amateur and unprofessional individuals and groups to come in and make a sham of legitimate science. Any activity witnessed or recorded at the site from hereafter will be completely unsuitable for detailed analysis due to an exponential cascade of contaminations proving, once again, that the modern day business of ghost hunting is less about the science and more about the fame.

I look forward to laughing at the first bit of "evidence" put forth by those who stay at the Winchester House.

When Fantasy Meets Reality: The Conjuring
August 2013

Another old-school horror movie hit the theatres last month called *The Conjuring*, and it's doing remarkably well as far as the horror genre goes; but unlike many flicks that are cookie-cutter yarns using the same tired formula, this one is–at least in part–based on actual events involving real people and two iconic pioneers in paranormal research, Edward and Lorraine Warren.

Half marketing ploy, half respectful homage, the producers of *The Conjuring* hype the "based on a true story" aspect, but the names Ed and Lorraine Warren have been well known long before this movie's release to those who've made a career of the serious study of psychical research. They were ghost hunters before ghost hunting was cool, back when it was serious science. They were rock stars to budding scholars like myself.

This isn't the first film to be based on their work, either. Unless you were born yesterday then you've at least heard of–if not seen–the 1975 classic, *The Amityville Horror*, which led to ten more films.

When Lorraine realized that she had psychic abilities and that she could go into homes where people were having problems such as those in *The Conjuring*, she saw an incredible chance to use it to help people, and help she did in the years since.

In 1952, the Warrens founded the New England Society for Psychic Research, the oldest ghost-hunting group in New England. With well over 10,000 investigations in their storied career, they have authored numerous books about the paranormal and case studies of various hauntings. Many of their books grace my office library.

During the 1970s and '80s, the Warrens were part of some of the most prolific case studies in psychical research and leading contributors to the advancement of Electronic Voice Phenomenon techniques.

Sadly, Ed stepped across the Great Divide seven years ago, but Lorraine is alive, well, and full of much of the same adventurous spark that made them legends in the paranormal field.

She says that Ed would agree that the haunting and possession depicted in *The Conjuring* was one of the most extreme cases they'd ever witnessed.

As far as the new film is concerned, it depicts– admittedly in typical sensationalistic Hollywood fashion– the story of the Perrons as they dealt with both benign and malicious spirits in their Rhode Island farmhouse with their five girls in 1971.

Warren says that the movie does a "pretty good job" at keeping the storyline close to what really happened, "I can remember the places where it was very bad such as the dirt cellar [in the Perron home]. I can remember my husband going down the stairs and there was a professor from a university in New Haven, Connecticut who wanted to see what was happening in the home. When I came a few minutes later, Ed signaled me to go upstairs. When I got to the top and I looked in this room and it was all dark and this grotesque face was in there and I made the sign of the cross in the air and said, 'In the name of Jesus Christ I command you to leave and go back to where you came from.' That was a bad case."

The Perrons themselves star in some of the promotional materials of *The Conjuring*, which seems to add truth that the happenings were true. "Because I was the youngest and the most vulnerable, I was approached more than anyone, and I actually had a relationship with that (ghostly) boy," April Perron says in one trailer.

Many of their cases have been debated over the years, especially the Amityville case–that many say was simply made up.

The president of the New England Skeptical Society, Steven Novella, doubts the story and told *USA Today* that "there is absolutely no reason to believe there is any legitimacy" to the Warren's reports on the Perron case.

Andrea Perron, in return, responded that *The Conjuring* "is a fair reflection of the chaos and danger we faced at the farm," and has stated "there are liberties taken and a few discrepancies, but overall, it is what it claims to be–based on a true story, believe it or not."

Hey, I'll be the first to stand up and say that a dose of honest skepticism is good. At the end of the day it's important to keep two things in mind: *The Conjuring* is a movie; as such, its primary goal is to entertain and make money. Therefore, view it with a light heart. Although the events in the film are glamorized with studio special effects, they are based on actual events that are meant to inspire and encourage debate, research, and–above all else– foster an open mind about a universe that we are only beginning to understand.

The Science of Paranormal Research

Aside from the varied forms of paranormal activity, the most important aspect to understanding the phenomena is to learn about natural science and how the environment impacts an investigation.

Environmental Factors of Ghost Hunting
June 2011

We've covered a lot of events, phenomenon, and terminology associated with ghost hunting over the months. Unfortunately, for many in the mainstream sciences these still amount to fringe science; so I thought it time to touch on some of the principles of environmental and natural science to show how they apply to paranormal research. For all the advances we've made in science, there is much about our physical world, let alone the spiritual and psychological, that we have yet to understand.

Many parapsychological theories propose that hauntings are caused by the environment's ability to capture or "record" events; and that the spirit, once free of its corporeal form, is able to exist outside the physical realm to manipulate and communicate with the material world at will. If the flashpoint of an event (the initial occurrence or catalyst) is charged with enough emotional and/or physical energy the event can be imprinted into the location.

Cultures around the world and through the ages view the world and attempt to classify its characteristics based on their own unique outlooks and beliefs. Science is no different. It attempts to understand through experimentation and research rather than through folklore and faith.

First is the theory of the spirit as energy. Electricity is a building block of life–the driving force of the central nervous system that provides us with movement, thought, memory, and sensory perception. If we are indeed beings of energy, then upon death of the physical body we simply revert back to a state of pure energy.

When dealing with paranormal theories, science has the wonderfully ambiguous term "natural." Natural refers to any real phenomenon that appears ghostly but is defined as being created by a "scientifically unknown property of the present nature." Simply put, science dodges the paranormal implications by classifying the things its instruments and methodology can't explain as "natural" and shelves them for later review and study. Think of this as the scientific equivalent of the saying "God works in mysterious ways."

Polarity is a term with both scientific and cultural meanings and connotations. It is the concept that everything that exists–natural, spiritual, or otherwise–has an exact equal but opposite companion.

Many of the tools and theories in parapsychology work on the premise that paranormal activity isn't just beyond the normal, but is the exact opposite of normal. The philosophical undertone to this thought is worthy of further analysis and discussion, I think.

Science regards the worlds of the physical body, the spiritual, and the psychological as three completely separate and mutually exclusive segments of existence. Energy is movement caused by either mechanical or chemical reactions. Electricity is, therefore, just a form of energy with no sentience or will of its own. It simply follows a prescribed and predictable set of variables; so when forms of energy present incidents that fall outside the rigid expectations of behavior, the phenomena are referred to in the aforementioned definition of a "natural" phenomenon.

The primeval worldview puts all matter and thought in the universe into one of five "energies" or "elements". Air, the first element of alchemical tradition, is the essence of intuition and learning; element of the east and the nature of the mind. Its modern-day counterpart would be the gaseous state of matter. The second, Fire, is the essence of purification and change; the element of the south and the nature of the will. Its present-day brother would be plasma. Water is the essence of love and fertility; the element of the west and the nature of emotions. Its physical properties are that of the liquid state of matter. Earth is the essence of grounding and stability; element of the north and the nature of balance. Its equivalent is the solid state of matter. The fifth element refers to all things spiritual: thought, emotion, faith, reflection, ideas, and inspiration. All five elements work together and in harmony with one another.

If we follow the five-element worldview, then the cardinal compass directions of north, east, south, and west have an impact on the investigation of a haunted location. Therefore, not only will your compass point out instances of EM fields, but the directions things travel, or the directions in which certain rooms or that where objects are situated have meaning in the general design of paranormal activity.

Spirits may be able to manifest within the full spectrum of energies due to their shedding of physical form, but because of our inability to see these spectrums with the naked eye we must rely on equipment designed specifically to monitor and document those ranges.

In order for a spirit to manifest itself or manipulate objects it must draw on the energy present in a location. This exchange causes the physical environment to be affected in quantifiable way.

The normal state of the surrounding air temperature is called ambient temperature; and, theoretically, hot and cold spots are areas where a higher than average temperature inconsistent with the environment is recorded along with paranormal events.

Electromagnetic energy is a hybrid of electrical and magnetic fields that binds nature and surrounds the planet. Spirit energies generate energy contained in the electromagnetic spectrum. Many groups say that high sensitivity to EMF manifests in nausea, paranoia, and hallucinations but there are no official studies conducted that prove EMF levels of any type have these effects. In fact ANY residential setting has a significant EMF field.

One thing I'd like to make note of is that Microwave radiation often gives false readings to EMF devices.

The Sun and Moon provide several states which affect a paranormal investigation, but for time constraints I will discuss these next time, as whittling them down to fit into a single article would not do them justice.

Environmental Factors of Ghost Hunting: The Moon
July 2011

We left off discussing some of the mainstream scientific and environmental factors that contribute to paranormal research and ghost hunting. I touched on many of those topics as best I could and ended with a mention of how the Moon affects investigations. We've got a whole new month and column to stretch out our minds and comfortably talk about this important aspect of our otherworldly stroll.

The Moon has played an important role in spiritual and psychological matters for as long as there's been recorded history; many modern terms, such as lunacy, derive from the effects that the Moon is claimed to have on our psyches.

As the lunar cycle waxes to full, incidents of psychotic behavior, violence, and crime seem to escalate; the phase of the New Moon also seems correlated to a rash of abnormal behavior. Current understanding of human psychology and physiology supports the observation that the Moon can exert significant influence on the human mind.

But is there really any scientific support for such claims?

Well, it so happens that this pattern has been proven in a variety of studies: The Journal of Clinical Psychiatry

(1978), a 1987 survey, and a study at the University of New Orleans (1995) are a few examples.

Statistically, the studies found that psychiatric admissions actually drop at the New Moon, but that they increase during the first quarter; homicides, suicides, aggravated assaults, and fatal traffic accidents all increasing as the Full Moon arrives.

As a professional in the psychology field, I find this interesting. If the Full Moon exerts an influence on rising ocean tides and also an increase in erratic human behavior, and psychiatric admissions drop at New Moons, that supports a hypothesis that the gravitational pull of the Full Moon is counteracted by the equal but opposite effect of the New Moon having an adverse effect on the chemical imbalance in the brain, perhaps due to a reduction in the gravitational pressure exerted on the fluids within the body.

It is well established in science that the Moon's gravity is the cause of the ocean tides and affects many different phenomena in weather and nature. The Earth is mostly water; but the human body is also made up of mostly water to the same percentages. So if the Moon affects the planet due to its abundance of water, why not the creatures on the planet made up of the same materials?

As long as we're discussing the Moon's effect on water, it may be interesting to note that sites with poltergeist activity report unexplained drops or small puddles of water in the room, and in general more poltergeist activity is reported near water; many poltergeist reports also involve kitchens. Several of the best EVP recordings and other captured evidence for hauntings occur near streams or in close proximity to groundwater. Early studies suggested that running water may generate a frequency that renders some people more sensitive to psychic phenomena.

Some scientists argue that the lunar effect on gravitational forces doesn't just affect the tides, but are linked to earthquakes as well. A study of 2000 earthquakes demonstrated that they occurred when tidal forces where ` over the epicenter of the earthquake. The study showed that there seems to be a lunar trigger. Another study in 1978 found that the Moon triggers particle flow, which disturbs the Earth's magnetic field. Eclipses are known to have an effect on the conductivity of the atmosphere, especially in the E-region, which in turn affects the Earth's geomagnetic field–the largest effects occurring in the morning around the time of equinoxes. A 7.4 magnitude earthquake in Japan occurred on the same day of the joint winter solstice/eclipse on December 21, 2010.

So if a major parapsychological theory contends that paranormal activity is tied to fluctuations in the geomagnetic field of the planet, then it stands to reason that investigations can benefit from close association with lunar events to produce effective results.

Paranormal activity often increases on the anniversary of death, especially if the Moon was full or new on the date the person died. A possible cause of this is due to the gravitational forces being increased and therefore providing additional energy to bridge the gap between the different planes of existence. Whether or not the person's death correlates to a Full or New Moon, it's wise to check for increased activity on those nights each month, or on whichever moon phase was nearest the time of death.

Paranormal activity is often easily influenced by subtle factors so it's possible that the Moon has a greater effect on them now that they exist in the non-physical realms.

Some researchers have noticed slightly better statistical results during investigations held during the New or Full Moon.

In addition to the effects of the Moon on spirits, researchers can have greater psychic sensitivity during the Full and/or New Moons. The Full Moon has been shown to affect the physical body, including menstrual cycles (which in turn affects the body mentally and physically); whereas the New Moon is related to increased influence on mental states.

There is definitely a lot of room for speculation about lunar cycles and increased paranormal activity, or perhaps more awareness of it.

So, have any of you noticed an increase or decrease in activity during certain Moon phases? Is there any credit to these theories, or is it all just lunacy? Let me know what you think.

Environmental Factors of Ghost Hunting: The Sun
August 2011

The Sun can provide several states which affect a paranormal investigation, and these must be taken into account when conducting a serious analysis of parapsychological case studies.

The Earth is not the only body in our solar system to create a magnetic field; the Sun creates its own magnetic field that is dynamic, and can change location and intensity over time. The Sun also emits X-Rays, which are relatively high-energy photons. A stream of these photons has a very effective penetrating power.

When looking for paranormal activity, we know that the relationship between magnetism and electricity is explored through the use of an EMF detector to look for unexplainable spikes in the fields surrounding an area.

The first step to how these relate to solar flares is to look at what a sunspot is.

A sunspot is an area of the Sun that is cooler than the surrounding area due to a stronger magnetic field which doesn't allow for the transfer of heat. This magnetic field forms below the surface, and extends out to the corona of the Sun appearing as a darker colored spot on its surface.

During a sunspot, a solar flare can occur when plasma interacts with the magnetic fields of the sunspot and bursts outward. During these flares energetic particles, x-rays, and magnetic fields affect Earth as a geomagnetic storm.

The Earth's magnetic field, called the magnetosphere, acts as a protective shield for the planet from stray magnetic fields and energetic particles. During the solar flares, when the plasma that is emitted from the Sun comes into contact with the magnetosphere, the plasma disturbs it, and is displayed as auroras–such as the Aurora Borealis and Aurora Astralis; otherwise known as the Northern and Southern lights.

During the bombardment of these fields, radio transmissions and power grids may be affected–usually as surges in energy. The extra magnetic forces are then at the disposal of paranormal entities and may allow for increased activity.

Sunspots are also noted to run in a cycle of frequency of visibility increasing and decreasing over time. These cycles run an average 10.8 years.

This may have a relationship to places of paranormal activity that seem to fade. If these places were fueled by increased magnetic fields from sunspot activity, it is possible that as the cycle decreases, then the catalyst for the paranormal activity decreases as well. This would mean that an area of high paranormal activity may die off and then return at a later time.

Fortunately, most of these particles are blocked by the magnetosphere and filtered by the atmosphere. It is theorized that ghosts and spirits might be able to use this constant shower of high energy particles as an energy source to manifest themselves, but solar activity isn't constant.

The Sun has periods of higher intensity emissions at certain times more than others. Scientists have determined the Sun goes through a 20-year period of generally high solar activity, followed by 20 years of relatively lower activity, and then the cycle starts again. The Earth is currently in the middle of a 20 year up-cycle in solar activity.

When solar X-rays are in flux it is referred to as an *Active* burst. Scientists classify these according to their x-ray brightness. There are 3 categories: C-class flares, M-class flares, and X-class flares. C-class flares are small with few noticeable consequences here on Earth. M-class flares are medium-sized and can cause brief radio blackouts that affect Earth's Polar Regions; minor radiation storms sometimes follow an M-class flare. X-class flares are major events that can trigger planet-wide radio blackouts and long-lasting radiation storms. Each category has nine subdivisions ranging from C1 to C9, M1 to M9, and X1 to X9 respectively.

A *Mega Flare* describes when an unprecedented solar x-ray event has occurred.

The X-ray Solar Status Monitor is the scientific monitoring of solar x-ray activities on the sun.

A Solar Terrestrial Activity Report is a graphed comparison chart documenting solar cycles. The report includes info on sunspot numbers, solar fluctuations, and indexing for Earth's geomagnetic fields. Increases in paranormal phenomena can be checked against this report to confirm if geomagnetic fields have increased. The figure below shows a series of solar flares recorded by NOAA satellites in July 2000:

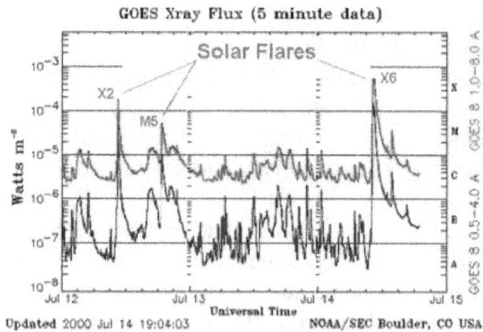

GOES Xray Flux (5 minute data)

There is an awesome freeware program I use called Ghost Weather Station which easily and effectively collects a lot of this data for quick recording while on site;

by Jonathan Moore, it is available for download at wvghosts.com, so don't get tricked by sites that ask for payment!

This tiny program provides (with a working internet connection) the current lunar and weather conditions of an area including the Moon phases. It will present Moon angle, percent of lumination, Julian Days, and lunation number; as well as the current Solar X-Ray and Geomagnetic Field status of the area after inputting the postal code assigned to the location.

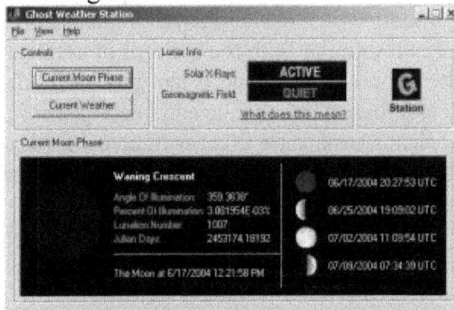

For a free program this can't be beat. I strongly suggest making a donation and downloading the program.

So, to sum it up–by noting solar activity during investigations we can determine whether or not there is, in fact, a clear correlation between increases in paranormal activity and solar activity.

Combining all of this information, along with the previous two installments on the Moon and other factors, we can clearly see that there is a wealth of accepted mainstream scientific fact that can be applied to paranormal research. It isn't all just smoke and mirrors, or capturing anomalies on film or audio.

By taking these factors into account when you conduct your investigations, you not only have a means of backing up your findings with irrefutable science, but you bring a level of respect and class to yourselves and the field, thus setting yourself apart from the thrill seekers and amateurs.

The Embodied Mind:
Altered States of Consciousness and Ghost Hunting
March 2011

We live in a time when science has eroded traditional spiritual beliefs so thoroughly that they are mutually exclusive for most people. This modern paradigm of the mind, body, and spirit being completely separate and unaffected by each other has warped our perceptions of the universe around us; but we need a sense of spiritual meaning in both a physical and psychological sense.

We're all familiar with the concept of mind over matter, the belief that the mind is more powerful than the body and can thus affect change and healing within the body. Medical journals are filled with colorful anecdotes and clinical studies involving placebos and faith-based healing.

Dr. Charles T. Tart is an American psychologist and parapsychologist known for his work on the nature of consciousness (particularly altered states of consciousness), as one of the founders of the field of transpersonal psychology, and for his research in scientific parapsychology. His first books, *Altered States of Consciousness* (1969) and *Transpersonal Psychologies* (1975), became widely-used texts that were instrumental in allowing these areas to become part of modern psychology.

Dr. Tart brings modern knowledge of parapsychology, altered states of consciousness, and transpersonal psychology together to show that, when properly applied, the scientific method actually gives us a strong basis for seeing ourselves as genuinely spiritual beings embodied in a marvelous body and nervous system. Such knowledge provides a needed basis for a general healing of body and mind.

But what can these theories show us as it relates to paranormal research and ghost hunting?

The term "Altered State" refers to any state of consciousness that is different from "normal" states of waking or sleeping. Altered states of consciousness include hypnosis, trance, ecstasy, psychedelic, and meditative experience; and do not necessarily have paranormal or dissociative features.

Let's look at how an altered state of consciousness shapes our subjective paranormal experience. When we accept the possibility of another plane of existence, we gain access to an entire level of perception. After all, seeing is believing, right?

Our conscious thoughts, feelings, and perceptions influence our interpretation of the things we take in. What we see, hear, and feel is influenced as much by our world view as much as it is by the sharpness of our body's reflexes.

There are no shortage of meditation techniques out there for centering and calming the body. No serious scientific data is going to be collected if your investigators are as giddy as school girls at a pep rally. Even seasoned investigators get wrapped up in the excitement of an investigation, especially when there is heightened activity. But by employing some simple meditations we can calm that excitement, while at the same time opening our minds, our bodies, and our spirits to perceiving more of the world around us beyond those related to the five physical senses.

Once your investigators are calm and centered they'll be more acute to the environment. But if we go beyond the person, into a deeper level of consciousness, an entire array of tools opens up to us as paranormal researchers. Many groups, including my own, have done experiments with Automatic Writing. This is writing done in a dissociated or altered state of awareness that is attributed to discarnate beings manipulating the writing tool of the subject in order to communicate while leaving the writer unaware of what is being written, or that any writing is taking place at all. One thing I'd like to point out is that auto writing has been blasted by skeptics for its highly subjective nature, but a true auto writing session results in a handwriting style that is markedly different from that of the subject. For this reason, someone who is ambidextrous should not be the one chosen as the subject in order to neutralize that x-factor from the experiment.

Scrying is an ancient technique of gazing into an object in order to achieve a heightened sense of awareness or altered state of consciousness. No, gazing into a crystal ball will not allow you to see clear visions like a movie clip. What it does is focus the eyes and mind onto a single object to facilitate a transition into an altered state. We can use this ancient technique in several ways using modern ghost hunting equipment to enhance our investigations.

The technique of mirror gazing is a simple one. Focus a video camera onto a full-length mirror. It's best to use a camera that has a large fold-out viewer, but if you have a camera with a single eyepiece that's fine. Make sure, either way, that you are sitting perfectly still for the duration of the experiment. The goal here is to use the mirror to reflect the camera lens back in upon itself into infinity. This is the focal point. As you conduct your spirit communication, any images that you see will also be recorded in real time and can be analyzed later for validity and clarification. A suggestion is to also use an Ion Air Purifier to spread positively-charged ions into the air.

Explore different techniques and equipment. If you've conducted your own experiments with altered states of consciousness, or have a personal experience, please share them.

"The Perception of Believing"
August 2012

It is often said that seeing is believing. That is to say that there is no definitive proof that something exists–that something is real–until it is seen or heard by the eyes and ears of a human being.

That seems a bit arrogant in its base. It is scientific fact that animals, such as dogs, sense and perceive things in our environment that are physically beyond our human limitations and comprehension, but that doesn't make it any less real.

Or is it that believing is seeing? That one cannot have a sense of something's existence unless they first believe in the possibility of it, thereby opening up their mind to having the experience.

As a psychology scholar I have had a very keen interest in the psychology of perception because the perceptual process is what allows us to experience the world around us; and it makes the human condition an enjoyable personal story.

No one's world experience is exactly the same. It is the end result of the beliefs, education, behaviors, and witnessed events specific to that individual; and as such, it is constantly evolving and changing.

No two people can have exactly the same set of these variables outside of exact clones programmed with the very same set of memory engrams, psychological profile, and physiology. But the moment the two are created they are no longer exactly the same because their view of the world, at least in the sense of visual acuity and position, is no longer the same.

Two friends can witness the same event, say for example a car accident, but remember the details in different ways, or even in a different order. Their proximity, field of view, emotional state, and memories all can play key factors in how they recall the incident.

The belief in, or the experiencing of paranormal activity may be the result of a coping mechanism within the perceiver to make sense of the witnessed phenomenon.

That's not to say that the experience wasn't real. In the mind of the perceiver it was–or is, for all intents and purposes–a very real event; but that doesn't necessarily translate to empirical evidence in the eyes of science.

That's why personal experiences and anecdotes related to psychical research are never used when determining the validity of claims of paranormal activity. These are by their very definition subjective rather than objective experiences. If it can't be quantified or confirmed by an impartial third party (i.e. scientific data collection) then it didn't happen.

It has been suggested by Joe Nickell, senior investigator for the Skeptical Inquirer, that there is no proof of life after death or other claimed paranormal phenomenon.

Using leprechauns as an example, he points to the lack of scientific evidence that leprechauns exist beyond the realm of cultural folktales because most people don't have any personal motive for believing in them or in seeing them.

William Butler Yeats would have retorted, "The world is full of magic things, patiently waiting for our senses to grow sharper."

"We also have no scientific proof that ghosts and extraterrestrials exist," he says, "yet most people do believe they're real." This is because of the promise of something more–something greater than the current human condition. Believing that we can go on after death or the belief that we're not alone in the universe has specific emotional incentives for most people.

The questions involving the perception of paranormal activity are as vast and as complex as the answers themselves. Entire studies have been conducted and full-length works have been written exploring both sides of the debate. But, as always, I open the floor to you, my fellow explorers. Is it that seeing is believing, or is it that believing is seeing? Perhaps it is a simple matter of seeing what we want to see.

There are things about the physical universe that we know, and there are a greater number of things that we do not know, and between them both is a door that swings both ways waiting for us to have the courage to step through.

"The Minefield between
Paranormal Belief and Religion"
May 2013

We've covered how to blend spiritual work into paranormal investigations in one of the earlier series of this column; and how the psychology of perception shapes our paranormal experiences; as well as a plethora of discussions on how scientific principles and methods apply to paranormal research.

Through it all, we've seen that the realm of parapsychology is a very unique and complex field with theories that span the entire spectrum of human experience–from philosophy and spirituality to environmental science and physics.

Recent surveys suggest that more than 67% of Americans hold some sort of belief in paranormal activity, and that number is expected to increase according to research by sociologists and other scholars. That interest is, of course, driven by entertainment on both the large and small screen from the likes of the *Paranormal Activity* movies and ghost hunting "reality" shows.

More than half of Americans also believe in the existence of extraterrestrial life–or at least the probability of it. This is largely due to the advancement of technology capable of searching the universe for planets in the "habitable zone," those capable of either creating or sustaining life in even its most basic forms such as those recently announced by NASA.

Statistically speaking, those who express a belief or personal experience in paranormal activity are not the odd ones out anymore as the cultural landscape of America changes. It used to be that to say you had a ghost in your house was met by a raised eyebrow and a call to the men in the white coats. Now it is met by fascination. By contrast, Europeans tend to think it odd if your house is the only one on the block not haunted.

The widening interest in the paranormal has taken on a model that sociologists point out goes back to the earliest hunter-gatherer societies. While men typically want to go out and capture something to prove its existence, women tend to want to use that information to improve themselves, become better people, and help others.

Sociologists Christopher Bader and F. Carson Mencken of Baylor University and Joseph Baker of East Tennessee State University report in their new book, *Paranormal America*, that unmarried and cohabiting individuals are far more likely to embrace the paranormal.

On a side note, they also found that members of the Republican Party were "significantly less interested" in the paranormal than Democrats or Independents.

At the core, though, they found that a conventional lifestyle and the firm grip of conformity were strong factors in non-believers while those with highly unconventional attitudes tended to look at explanations beyond the confines of mainstream thought.

Over the years, the most controversial–yet frequent–discussions that I've been part of on the subject revolve around the delicate minefield that exists between religion and the paranormal.

The groups most likely to remain on the side of non-belief are atheists, fundamentalist Christians, and Jews. In fact, the most committed of those individuals–those who attend services weekly–are among the least likely to hold paranormal beliefs. Those who believe the Bible is the literal word of God are also highly unlikely to entertain the existence of ghosts, clairvoyance, and other related phenomena.

However, a new generation of spirituality seekers is opening their minds–and their wallets–to the belief in the paranormal and other phenomena not easily discussed among the more mainstream of religions.

There are some stark contrasts among the debates regarding the relationship between religion and the paranormal. Some suggest that those whose beliefs are outside that of mainstream religion embrace the paranormal as a substitute for a personal philosophy. Others say that religious individuals, or those already open to transcendent ideas and godly abilities, would be more likely to hold paranormal beliefs.

Most people tend to occupy the middle of the spectrum, though. These individuals, who are not regular attendees of services, have an interest in religion but maintain a greater acceptance in the paranormal. Belief in paranormal topics was found, for example, to be at its highest among people with more liberal views of the Bible.

What is particularly interesting is that the very core of most religions contains concepts and imagery that is very much paranormal and unexplainable. Many shrug off ghosts as being the stuff of fantasy, yet hold an adamant belief in angels, demons, and other spirits who carry out the will of God and the Devil.

What of clairvoyance, telepathy, and other related phenomena? Was Joan of Arc crazy or did she really have a telepathic link with God? Muhammad received his visions through dreams with the angel Gabriel. The very basis of religion, it seems, demands a belief in the paranormal in order to make sense of the stories that have been passed down since antiquity of angels and demons walking among us and carrying out magickal acts that defy logic and science.

Perhaps the answer lies out there, in the middle of that minefield that lies between religion and the paranormal. Are you brave enough to unlearn what you have learned and step into the unknown? See you out there.

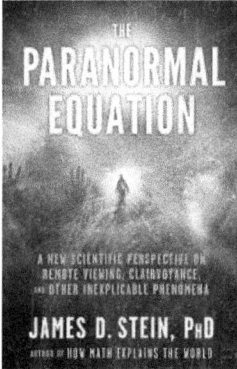

If you're reading this, then that means we've survived the apocalypse. Whew...what a relief that is. I really wasn't looking forward to poor hygiene, scavenging for food and energy, and martial law in a wasteland of zombies and *Mad Max* wannabes.

So now that the doomsday prophesies have failed to come true (yet again), and we chuckle at the silliness of the paranoia that has raided the airwaves over the last few years, we can return to the business of science and the quest for answers to questions that have yet to be fully understood and explained.

I'm a believer in the theories and principles of parapsychology–any of the long-time readers of this column know that. What they also know is that I don't just blindly believe anything I read on the subject, or think that trolling around cemeteries in the dark constitutes valid scientific method. I look at things with a believer's heart, but analyze them with a skeptic's mind.

James Stein is a Professor of Mathematics at California State University whose previous books include *The Right Decision: A Mathematician Reveals the Secrets of Decision Theory* and *How Math Explains the World*. In a book that was released last week, *The Paranormal Equation*, Stein provides a new scientific perspective the realm of supernatural phenomena–something many consider to be not only unscientific, but inexplicable.

He travels through the various definitions involved by cracking open a Merriam-Webster dictionary and describing supernatural as "of or relating to an order of existence beyond the visible observable universe;

especially: of or relating to God or a god, demigod, spirit, or devil."

But science doesn't restrict to the observable or visible universe. If it did, then subatomic particles, bacteria, radio, and x-rays would all be considered supernatural phenomena.

He further explains that the supernatural "depart[s] from what is usual or normal especially so as to appear to transcend the laws of nature."

Ah-ha. Now this is certainly within the ability of science to explore.

Transcendent laws have been implied by mathematicians in recent decades, such as Godel's 80 year-old Incompleteness Theorem, which establishes inherent limitations of doing arithmetic due to uncertainties.

Stein takes issue with combining words like 'observable' and 'visible' and states that once natural laws are uncovered, the events witnessed cease to be considered supernatural and lose their apparent mystery. So if science were able to adequately explain apparitions they would then cease to be seen as supernatural and henceforth be seen as natural.

Stein delves into detail in an attempt to explain that just because things happen by chance, it doesn't have to mean they should automatically be labeled as supernatural. A lot of parapsychology research, for example, points to chance when it comes to such things as ESP or clairvoyance–especially when those experiments are reproduced.

Many so-called 'paranormal' researchers and ghost hunters have no understanding of chance at all–or science for that matter.

Stein also explains that chance is actually incorporated into many of the cornerstone theories of physics. Given enough time, the most unlikely of things can and will happen as long as they are simply unlikely and not downright impossible–such as the ability of ice to form in warm water.

Stein is convinced, and successfully persuades the reader to believe, that paranormal phenomena must exist under the hypothesis that the Universe is infinite.

It should also stand that just because something is strange or unexplained does not mean that it is proof of the existence of paranormal activity or supernatural phenomena, either.

Gravity is an unseen force once completely misunderstood that is now accepted as scientific fact. The list of similar examples has grown exponentially in the last half century, so who's to say what other theories studied by parapsychologists will eventually be proven and accepted by mainstream science?

"What Does Science Have to Fear from
Parapsychology?"
February 2013

Last month we discussed some interesting points about science and the unanswered questions raised by parapsychology–a noted professor even attempted to explain some of the principles studied by researchers through the application of mathematical reasoning.

However, the underlying question remains: just why, more than 100 years later, does mainstream science still laugh off members of the field as one would their wacky uncle at Thanksgiving dinner?

The clichéd question many skeptics like to use is that if psychics are real, why do none of them ever win the lottery?

In *Science & Psychic Phenomena: The Fall of the House of Skeptics*, Chris Carter contends that psychic powers have not been conclusively proven because scientists are either blissfully ignorant of the available research or they simply refuse to take what is published seriously.

Why? Just what do they have to fear from opening their minds to new possibilities? Isn't that what science is supposed to be all about?

The hardline opponents most often refuse to acknowledge the existence of paranormal phenomenon because to do otherwise would cause a paradigm shift in how the universe is viewed through the collective mind of humanity's adolescent intellectual level.

Skeptics and scientists are as deeply committed to maintaining the status quo of their belief structure as anyone else is. There's a psychological defense–a certain comfort level–involved in the avoidance of a possibility, however improbable, that one could be wrong.

This belief results in closed minds, but this adherence to the status quo only applies to theories presented by members of the study of parapsychology; in mainstream scientific circles, a theory that could shatter long-standing laws of physics isn't laughed off so readily.

In September 2011, news shot around the world that Italian physicists had measured particles traveling faster than light. If validated, it would have violated the fundamental laws of physics and completely changed our understanding of the Universe.

The reaction from the scientific community to the news was not one of ignorance; they didn't brand the scientists hoaxers and quacks; they didn't shout, "Blasphemy, ye witch! Burn!" Instead, they did what any reputable scientist does when confronted with such a challenge: they took a closer look and tried to replicate the research. As it turned out, the anomaly was caused by measurement and mechanical errors in the original experiment.

Such scientific brotherhood is not generally the case when it comes to theories presented by psychical research; every now and then, though, an exception is made.

A study published in a 2011 edition of *The Journal of Personality and Social Psychology,* by Cornell professor Daryl J. Bem, claimed to have found strong evidence for the existence of psychic powers such as ESP; it quickly made headlines around the academic world for its implication that psychic powers had been, finally, scientifically proven.

Bem's claim wasn't ridiculed or ignored; instead it was taken seriously and tested by scientific researchers; replication, being the benchmark of any valid scientific research, should be able to replicate the findings. If not, then the findings could be summarily written off as random variations and errors.

So, a team of researchers set out to replicate Bem's experiment and found no evidence for psychic powers. Their results were published and Bem publically acknowledged that the findings did not support his claims and wrote that the researchers had "made a competent, good-faith effort to replicate the results of one of my experiments on precognition."

While this was a heartwarming exception to the "black-sheep-of-the-family rule," the reason that research looking into psychic powers and hauntings are rejected by the scientific community is simply because there hasn't been that jaw-dropping discovery to make the world scream in unison, "Holy crap! Sorry, dude, we were wrong. Our bad." Yes, some major players in the scientific community do talk like that—they aren't all stiffs with pretentious accents and bow ties.

The evidence for psychical phenomenon, like anything else, rests solely on its own merits. Better research follows with better evidence; and as technology readily catches up with the theories that began over a century ago, the answers may finally come. So we patiently await that Nobel Prize moment that changes the course of human understanding. It may not be tomorrow, next week, or next year; but the possibility, however slight, should not rule out the quest for it as absurd or wasteful.

There's no reason that science should fear or dismiss the study of the unknown. That's the one thing that unites us all—our insatiable desire to question everything and the adventure essential to the quest.

"Where, Oh Where, Has the Science Gone?"
January 2014

There was a time, about 40 years ago, when paranormal research was fueled more by a passion for science and understanding than for fame and glory. It was a time when some of the world's most prestigious universities were home to respected parapsychology departments conducting cutting-edge research and publishing their theories and findings in the world's top academic journals, where they were discussed and graded on their methodology and scholastic merit. It was a great time for the field. Well, before Hollywood and groups using poor techniques and duct-taped equipment stumbled around with night vision cameras looking for ratings rather than answers.

Somewhere–lost in the shadows of the scripted "results", camera magic, and glitz of reality television's idea of what paranormal research is–the honest, academic exploration of parapsychology has all but disappeared from the public eye, vanishing into the mists like a Gray Lady on a windy autumn night. Or has it?

For several decades, Duke University ran one of the largest centers for paranormal research in the world: The Duke Parapsychology Laboratory, established in 1935 as part of the college's main psychology department. Other programs at Stanford and UCLA followed, but the academic environment was about to change.

Remember what happened to Dr. Venkman and company in "Ghostbusters" when they were kicked out? Those scenes were based on the real history of the Duke program when skeptics derided the field of parapsychology as pseudoscience and Duke ended its affiliation with the program. J.B. Rhine moved his labs across the street–literally–and continued the work off-campus that he and William McDougall started as the independent and privately-funded Rhine Research Center.

While it may be true that in recent years the field has become fragmented and underfunded, it still manages to hold on, even as the mainstream scientific community has once again ignored it. John Kruth, executive director of the Rhine says, "It hasn't gone anywhere. People have never stopped doing research in these areas, but the skeptic community is strong and vocal, and they're much better at working the media."

Kruth points to media-savvy skeptics such as James Randi for much of the academic community's attitude toward the field. I would have to agree with him, at least in part; but it's this level of blind skepticism that can't accept anything that does not fit into rigid definitions and worldviews. Anyone who dares look at other possibilities is immediately labeled a fraud and excommunicated via a round of laughter.

This is a mindset summed up in the words of Michael Shermer, editor of the quarterly journal *Skeptic*, and columnist for *Scientific American*: "It's [parapsychology] fallen into disuse due to the fact that there's just nothing there."

"Certainly there are fraudulent practitioners out there, and we're always watching for that," Kruth said. "It's like we have the frauds on one side and the debunkers on the other, and we're in the middle, still trying to do science."

Critics retort that parapsychology, as a field of scientific study, has a fundamental evidence problem. I would have to agree, but I think Shermer is wrong when he arrogantly claims that "parapsychology has been around for more than a century. [Yet] there's no research protocol that generates useful working hypotheses for other labs to test and develop into a model, and eventually a paradigm that becomes a field. It just isn't there."

At a recent presentation on the campus of Duke University, three parapsychology researchers presented results from their latest studies. The presentation topics: "Synesthesia, Time and the Geography of Anomalous Experiences"

"Synchronicity and Psi: A Controlled Comparison"
"The Bio-Energy Lab at The Rhine and The O.B.E. (Out of Body Experience) Project"

Other recent research has huge implications in both parapsychology and mainstream clinical psychology. Surely it's not your Vulcan mind-meld level of telepathy, but researchers came closer than ever to getting one mammal to read another mammal's mind.

A research team had wired together the brains of two rats, allowing them to transmit information between each other and cooperate. The results, detailed in the journal *Scientific Reports*, could help improve the design of neural-controlled prosthetic devices and perhaps even show that one day we could network brains as well as computers, or communicate by translating neural activity in the brain into electronic signals.

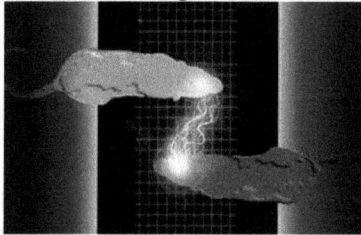

In the experiment, the Duke scientists trained two rats to press one of two levers when a particular light switched on. Next, they connected the animals' brains with tiny electrodes, each a fraction the size of a human hair, that linked the parts of the rats' brains that process motor signals. Rat #1 was the "encoder" while rat #2 was the "decoder." The first rat's job was to receive the visual cue to press the lever. If it got it right, it got a reward.

As the encoder rat did its task, the electrical activity in the encoder rat's brain was then translated into a signal and transmitted to the decoder rat. That rat would then press its own lever. For the second rat, though, there was no light cue to tell it which corresponding lever was correct. It could only go by the signal it received from the other rat.

It hit the correct lever an average of about 64% of the time- sometimes up to 72%, a result much greater than possible by chance. To confirm that this was an effect of the signals from the encoder rat's brain, the team gave the decoder rat the same stimulation–but this time from a computer–with similar results.

Another experiment tested whether the rat's brain could transmit information about touch. This time the rats were trained to put their nose through an opening and, using their whiskers, distinguish whether the opening was wide or narrow. For wide openings, the rats were taught to poke a computer port on their right. For narrow openings, they poked to the left.

Once trained, the rats were wired to each other. When the encoder rat poked the relevant port, the scientists recorded the brain activity and sent the signal to the decoder rat. The decoder chose the correct side, left or right, to poke 60-65% of the time.

These research models show promise for parapsychologists as they examine psi and other extra-sensory forms of communication.

Perhaps what researchers should do is employ a bit of psychology when publishing their findings to show just how much valid "research protocol and working hypotheses" there is in the field. By running their experiments with parapsychological theories in mind, and making no mention of that fact when publishing their experiments as typical, mainstream neuropsychology–only then may be able to see just how deep the animosity and arrogance truly is among the world's scientists. Remember that some of history's greatest minds were laughed at in their own time, only to be repeatedly proven correct decades and centuries later, especially once the correct level of technology came along.

In the meantime, enjoy this video showing the most detailed map to ever be drawn of the human brain: http://www.newscientist.com/article/dn23731-3d-map-of-human-brain-is-the-most-detailed-ever.html

Paranormal Communication
February 2014

To exist is to communicate. Bees in a hive, a pack of wolves on the hunt, two people in a coffee shop, even the phones we text each other with; effective communication is occurring everywhere around us every day and in various forms both organic and artificial.

Humans are quite egocentric when it comes to the physical world and their professed mastery of it. If they cannot hear it, see it, touch it, or understand it then it doesn't exist or is laughed off as the illogical ramblings of the foolish. A common, and clichéd, axiom is "if a tree falls and no one is around to hear it, does it make a sound?" This naïve reasoning asserts that since no human was around to hear it, then it didn't happen–that it is impossible to occur. The wildlife that scattered as the sound pierced the serenity might feel differently.

Humanity assumes that the only effective, real, communication between two sentient beings is in the verbal patterns of spoken language; but true communication occurs all around us every day without us even being remotely aware that it is happening- or how. Just because we do not readily perceive or understand it does not dismiss this fact.

To effectively communicate with the world around us means that we have to let go of this arrogant mentality and realize that all things are different in the way they speak to each other and to the environment around them. This is the central canon when learning to understand and communicate with the world on a deeper level.

Why is it that emotionally-connected people can know exactly what each other are thinking or feeling without a single word being spoken? A look from a beloved dog or cat can elicit intense understanding in the same way. These are forms of paranormal communication. When one knows how to listen, communication and understanding come easy.

We saw last month that parapsychology researchers have made interesting discoveries about the possibilities of telepathic communication between two beings; and we've previously covered the topic of Electronic Voice Phenomenon with some intriguing examples that I personally captured. (It should be noted that Instrumental Transcommunication (ITC) is the more general, accepted, term in parapsychological research for any variation of device and corresponding phenomenon collected.)

Many believe that communication with the dead is a scientific fact and experiment with a variety of techniques for spirit communication to provide evidence of the continuation of life after physical death. Many parapsychologists and ghost hunting groups agree, which is why it is such common practice to attempt communication with spirits using devices such as voice recorders and cameras.

Mainstream science has generally ignored ITC, finding the results less than credible, and references a percentage of recordings that turn out to be hoaxes created by frauds or pranksters. Many also regard the examples put forward by proponents as simply misinterpretations of natural phenomena, explained via a variety of known psychological and physical phenomena; the tendency of the human brain to recognize patterns in random stimuli and radio interference are respective examples. I am not arguing the validity of the recorded samples that I or others have collected over the years as absolute proof of spiritual communication nor am I completely debunking them but if anyone has sound, logical explanations for their origin then, please, enlighten us; if not, then take it with a grain of salt and an intellectually open mind.

Pareidolia and Apophenia are the most common basis for arguments against the legitimacy of ITC.

Auditory pareidolia is a situation created when the brain incorrectly interprets random patterns as being familiar patterns. In the case of ITC, it could result in an observer interpreting random noise on an audio recording as being the familiar sound of a human voice. The tendency for a "voice" heard in white noise recordings to be in a language understood well by those researching it, rather than in an unfamiliar language, has been cited as evidence of this; and a broad class of phenomena, referred to by author Joe Banks as "Rorschach Audio", has been described as a global explanation for all manifestations of ITC.

Apophenia is related to–but distinct from pareidolia-in that it is the spontaneous finding of connections or meaning in things that are random, unconnected, or meaningless; and has been put forward as a possible explanation.

Suppose, for example, that an English-speaking group such as Ghost Hunters International were in Germany on an investigation. Would it not seem logical that any samples recorded and cited be in German rather than English? If a group's team members are walking around an Italian castle asking questions in English how do they expect any possible spirit to understand the question, let alone respond correctly and in English? When they analyze the audio, are they listening for English and Italian words, or perhaps another language entirely?

I have developed a theory that attempts to explain Rorschach Audio.

Everything about life I learned from watching Star Trek. Humor me for a moment. Ever wonder how the intrepid crew of the Enterprise can travel the galaxy and always find alien cultures speaking perfect English? Well, they have an ingenious little device called a universal translator that puts the spoken words through a sound algorithm that searches for patterns and then correlates them to the correct English vernacular. Neat trick, I think. However, there's often a fine line between science fiction and reality. Modern language translation apps on smartphones are the best example of this new-found reality.

In Jungian psychology, there is a bedrock principle that part of the unconscious mind is common and accessible to a group, a society, or even all of humanity, which is the product of all the ancestral experiences of a people throughout time and shared by all, termed the Collective Unconscious. Based on this model–if, as many world cultures similarly believe, we are all related, and taking into account the scientific theory that we revert to pure energy after the death of the physical body–then it stands to reason that as we pass from the material world into the realm of pure energy we have access to all of the knowledge of our fellow man. Therefore, if one were Russian in physical life, then such a being would have knowledge of the English language after death and would thus be able to communicate freely and fluently with an English-speaking researcher conducting an ITC session.

It might not be too hard to imagine the possibility. It is a known fact that the human body creates, stores, and uses electrical energy. Our neurology is much like a computer's in that it transmits data in much the same way as a circuit board. This explains how ITC is possible because if, after death, we revert back to pure energy then that energy is quite capable of manipulating and being captured by recording devices that operate on the same principles.

The old-style method of using cassettes to communicate with spirits is a testable theory because those mediums rely on the creation of electromagnetic fields to imprint the data on the magnetic strips. This is also why spikes in EM fields are of major interest to researchers and ghost hunters because it hints that spirits must draw upon the energy in the environment in order to be discovered. The modern digital recorders are similarly valid because of the means by which they retain information.

In a recent conversation online, it was jested that if ghosts did exist, it was doubtful that they would be using Facebook as a means to communicate. Based on the points raised today I would have to disagree. The internet is a digital world, using electrical currents and electromagnetics to transmit data from one side of the globe to another in an instant, and is just one more avenue that a being of energy could manipulate in order to communicate—a literal "ghost in the machine" as it makes its voice heard.

It is said that we have two ears but one mouth because listening is twice as hard as speaking. This is a universal truth. If we simply learn to listen to the world around us in a different way, we just might hear things that we were previously unaware were there.

Quiet yourselves and listen carefully to the sounds of the universe. The truth, and understanding, are out there if we are simply willing to listen with an open mind.

Do We Need Parapsychology?
February 2012

When one speaks about a topic which is controversial it is important to understand the concept of a paradigm, or underlying worldview. It can be thought of as a framework of beliefs which are so taken for granted that most people are not even aware they have made any assumptions.

A paradigm helps us to make sense of the world around us. In terms of science, it not only determines what is true, but how truth itself is determined. There is an obvious "catch 22" to this: If one doesn't recognize the underlying assumptions made with a paradigm, it has the potential to limit our perception of the world, what we can discover, and how we can determine that knowledge.

The old paradigm, which many have held since the days of Descartes, states that the subjective and objective worlds are completely distinct, with no overlap. Subjective is "here, in the head," and objective is "there, out in the world." The Cartesian paradigm presupposes that there are objective ways to define and measure the fixed external world, which the followers of this paradigm would say is the only world that matters.

Writer and philosopher Elbert Hubbard (1857-1915) eloquently quipped that "the supernatural is the natural, just not yet understood."

The formal, scientific, study of paranormal phenomena began in 1882 with the foundation of the Society for Psychical Research in London, England. Early efforts attempted to dissociate psychical phenomena from the pop culture trend of Spiritualism and superstition, and to investigate mediums and their claims of evoking spirits or apparitions.

But 100 years later most people still think that paranormal research is either a group armed with night-vision tech, stumbling around buildings in the dark in search of ghosts and fame, or simply the study of any subject that is weird or bizarre (e.g. Bigfoot and

UFOs/aliens). Parapsychology is—and has always been—so much more than the former, and has nothing at all to do with the latter.

Paranormal research does *not* concern itself with UFOs, urban legends, vampires, witchcraft, or mythical creatures (a study known as cryptozoology).

What parapsychology *does* study is the seemingly abnormal qualities of the physical universe in a scientific quest to find order and meaning in life. It is the ultimate exploration of the human condition and the discovery of all that the brain is capable of becoming; some of these concepts the legendary Carl Jung touched on with his theories of the collective unconscious and synchronicity.

Many people inappropriately use it as a synonym for "paranormal investigators," such as when referencing the cast of *Ghost Hunters* or *Ghost Adventures*; moreover, parapsychologists have also been linked with "psychic" entertainers, magicians, and illusionists. Some self-proclaimed "psychic practitioners" even falsely claim to be parapsychologists, going so far as to wave about bogus doctoral credentials.

This is not to say that all psychics are that way. I am personally acquainted with a few very adept and talented psychics here in the Detroit area. Life, however, is rarely as glamorous as Hollywood portrays for them. At best they are ignored or written off as delusional; at worst they are harassed and fired from their places of employment. Often psychics are exploited by mainstream media for fluff pieces in October, and mocked by the same news organizations the other 11 months of the year.

There are the inevitable frauds, scammers, and crooks. This is an unfortunate truth, and a few bad apples have spoiled it for everyone else. It is inexcusable that these charlatans con money out of vulnerable and naïve people. This is why no respectable group ever charges for its services.

It should be noted that many parapsychologists take an empirical, data-oriented approach to psi phenomena.

However, some researchers regard the current findings of parapsychology as having a wide variety of important implications about the spiritual, physical, and psychological nature of humankind.

Parapsychology is fascinating because of the implications it places on society, science, and how we understand the very nature of existence. The study and theories related to psi phenomena are not suggesting that mainstream science is wrong, or arrogant. However, it does suggest that what science knows about the nature of the universe is incomplete–that the accepted limitations of human potential have been underestimated–that Western assumptions and philosophical beliefs about the separation of mind and body may be rash–and that religious assumptions about the divine nature of miracles might have been misinterpreted.

Physicists have an interest because of the proposition that we have a misunderstanding about space and time, and the transfer of energy and information. Biologists are interested because psi implies the existence of non-physical methods of sensing the world. Psychologists are interested in the theories regarding the nature of perception and memory. Philosophers are interested because psi phenomena specifically address many age-old philosophical debates concerning the role of the mind in the physical world, and the nature of the objective vs. the subjective. Theologians and the general public tend to be interested because personal psi experiences are often accompanied by feelings of profound, deep meaning.

A cornerstone of the current scientific worldview is that human consciousness is nothing more than a result of the functioning of brain, body, and nervous system. No matter how different the mind may seem from solid matter, it is generated solely by electrochemical functioning and so it is absolutely dependent on it. When the brain dies, so does consciousness. From this perspective, claims of the survival of bodily death and the resulting apparitions are mere wishful thinking. Furthermore, the limits of material functioning automatically determine the limits of mental functioning, thus ESP and PK are impossible, given the establishment's understanding of how the world works.

Still, psi phenomena have occurred in all cultures throughout history, and continue to occur; some of the reported phenomena have also been convincingly verified using scientific methods. Because psi seems to transcend the assumed limits of material functioning some interpret psi as supporting the idea that there is something more to the mind than just the firing of neurons and electrochemical reactions.

This "non-physical" aspect, which is not restricted by space or time, might survive bodily death. If so, there may be important truths contained in some spiritual ideas and practices.

The research in parapsychology may have implications for spiritual concepts but parapsychologists are not driven by some hidden spiritual agenda. Some critics of parapsychology seem to believe that all parapsychologists have hidden religious motives, and that they are really out to prove the existence of the soul. This argument is as absurd as claiming that all chemists have a secret agenda in alchemy, and the quest to attain riches by turning lead into gold.

Despite all its claims and advancements, there are just some things that mainstream science can't explain about the universe. Parapsychology really acts as the center of scientific doctrine and theory, with lines leading to and from every branch of the other sciences. Together they form an intricate web of knowledge and understanding that is only limited by the egotistical whimsy of those who think they know all there is to know about the nature of the universe based on their blind obedience to one limited train of thought.

A Ghost Hunter's Toolbox

A ghost hunting series must deal with all of the cool toys and innovations at our disposal, so this brief "ghost hunter's toolbox" focuses on the most common—and the most ridiculous—of equipment at our disposal.

The Harsh Truth About Ghost Boxes
January 2012

The various ghost hunting "reality" shows that plague the airwaves have given a great deal of attention lately to an amusing new line of gear that merges EMF, audio recorder, and K-II devices all into one unit; some even assert to turn this data into spoken words that they spout as proof of spirit contact. While in theory this sounds fantastic, in practice it's a very different, very sobering reality.

Not only are these devices laughable at best, but these "professional" ghost hunters are actually trying to pass off the data from these devices as legitimate evidence of paranormal activity. These devices are complete crap. Come on, folks- this is supposed to be science, not a scene from the transcript for *Ghostbusters III*.

There are numerous versions of these devices readily available for sale on eBay; and YouTube abounds with video clips of their supposed findings. It's no shock to learn that the fine "professionals" over at Travel Channel's *Ghost Adventures* swear by these toys. That, if anything, is proof enough to discredit these devices and their data.

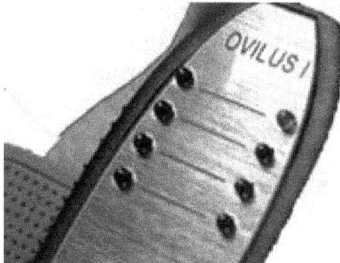

I first came across this type of device a few years ago when I heard of the Ovilus. Created by Bill Chappell of Digital Dowsing and appropriately labeled "for entertainment only," it claimed to translate EMF fluctuations into phonetic speech by converting the EMF readings into numbers, and then those numbers into words by sounding them out using text-to-speech algorithms via a vocabulary of 512 words. Various modes on the device include speech mode, which uses the environment to pick the words to say; phonetic mode, which uses the environment to create words phonetically; commutation mode, which uses speech mode and phonetic mode together; EMF Mode; yes/no mode, to ask questions and get yes or no answers (a digital Ouija board?); level mode, to watch the energy change in the environment; and dowsing mode, to work like a pair of dowsing rods.

All of this tech at your fingertips is powered by a common battery and equipped with a headphone jack, a recording jack with attenuated output, and something called the Paranormal Puck. The Puck is designed to aid in paranormal research and meant to be the "center" of investigation as a place to gather, log, track and maintain the data. It also watermarks data to prevent tampering. Users note that it can be "randomly repetitious" at times by stating selected words for every question asked and every environment investigated.

Ahem Really? Say it isn't so?

The first question that comes to mind is how can the inventor of this device possibly test the results? What evidence or reasons are the formulas based on? Whatever method he used to equate EM energy with words would have to start as an arbitrary guess. It would then need to be tested repeatedly to verify the results. In any case, this makes me think of the dog collars that supposedly turn barking patterns into words like "outside" and "water". Seems to me that this is just another example of wannabe researchers barking up the wrong tree.

The fine folks at Paranormal Research & Resource Society frequent their local Radio Shack for their "ghost boxes."

Known as the "Radio Shack hack," it was invented in 2007 by a retired electrical engineer. These are modified AM/FM radios that continuously scan the various bands to create white noise in the belief that entities can use the audio falloff from broadcasting stations to communicate.

One model, the 12-469, simply produces a clicking sound when scanning through the bands; other models are modified armband FM radios from the likes of Jensen that are common among joggers.

A man named Frank Sumption invented a version of the device after experimenting with software to record EVPs. His device would produce random voltage to create raw audio from an AM tuner, which was then amplified and filtered into an echo chamber for recording.

What makes these boxes unique in terms of EVP analysis is that because they were originally radios and equipped with external speakers the addition of the ability to record sounds, proponents say, can potentially lead to real-time two-way communication with the other side.

Not surprising, many users report that results of the ghost box are affected by the strength of the radio signals in the area; poor signal quality reduced the ability for spirits to make contact (*insert facepalm slap here*).

Furthermore, what conclusive proof do users have that the voices are indeed paranormal in nature and not simply the broadcasts of local stations? Depending on the atmospheric conditions one could even pick up a station from great distances. This is not unlike an experience I had with a CB radio some years back. While driving in the northern suburbs of Detroit one clear summer night I ended up in a chat with a trucker outside of Las Vegas, all because the atmospheric conditions made it possible.

Anyone with the latest generation of Smartphone can even download an app (often for free or a few bucks) that claims to do the same. Ghost Radar is one that comes to mind that I've personally come across. These are toys, nothing more. If that's your team's idea of science, stay at home and play Angry Birds instead.

I'm all for inventiveness, and I think some of the reasoning behind these devices has some merit; but these self-made devices are tainted by their very nature. No conclusive proof could ever possibly come from them unless the findings can be proven using other verifiable equipment as a control measure. As with the field of paranormal research itself, the tools and theories behind them need to go through extensive experimentation and testing to prove or disprove their validity for recording and measuring paranormal activity, let alone the resulting data that is collected by them.

One again we have the misguided practice of amateurs doing a disservice to and disrespecting science. I applaud those who invent these ghost boxes, as necessity is truly the mother of invention and I admire the resourcefulness involved; but I must condemn their inept notion that anyone with an intelligence greater than a radish take their findings seriously. The Ovilus and the various ghost boxes need to undergo years of intensive experimentation in various settings and controls to not only prove their worth, but decisively identify what sounds or readings mean exactly which words or phrases.

I've said it before, and I'll say it again—it doesn't matter how new or fancy the technology is, a tool in the hands of the unwitting is just a toy.

Science and Psychics–The Tech of Paranormal Research
November 2011

Most of the intellectual rhetoric thrown back and forth between skeptics and parapsychologists concerns the types of tools used during investigations; sometimes even those within the field of psychical research will argue among themselves for or against certain techniques and tools.

Since the field is one which attempts to quantify and classify phenomena that are, by definition, cultural, religious, and fundamentally unknown, it is somewhat acceptable to utilize devices and techniques of a more "mystical" nature. Many times the use of arcane devices and psychics can help lead the team to an area of interest or heightened activity, and sometimes even actual contact with the netherworld.

Once these devices or techniques have pointed the way to the presence of activity, the seasoned researcher will switch to more scientific instruments to document any possible activity. Unfortunately, the truth is that at the end of the day it doesn't matter what kind of personal experiences, thoughts, feelings, intuitions, or psychic imagery is collected, or by whom–if it can't be verified or quantified through impartial scientific measurement and documentation, then it technically never happened and just becomes yet another account in the mythos of a location's "ghost stories."

Tools have been modified and adopted from various sciences and applications over the years to measure and analyze data in a paranormal investigation. Some devices are used specifically to debunk phenomena and establish clear natural causes, while others have the purpose of capturing evidence–such as voice and video recorders. EMF detectors have a unique function of being used both for the debunking and the signifying of paranormal activity.

However, regardless of how expensive or scientific the tools, they are only as scientific as the person using it.

A team may boast about owning the most sophisticated thermometer available, but if members are using it as a barometer, the measurements are worthless; just as using a calculator doesn't make you a mathematician, using a Geiger counter doesn't make you a scientist. In the wrong hands the most accurate measurement device is nothing more than an expensive toy.

All paranormal research groups have their own unique procedures and instruments of choice. Some are religiously-based and use age-old tools and techniques; some consider themselves ultra-modern and use only the most expensive and scientific of equipment. Most groups, however, fall somewhere in the middle; and the tools, techniques, and even the very members come from a vast array of backgrounds, philosophies, and religions. The make-up of these groups and the tools they use are contingent on finances, personal preference, and practicality.

We're all familiar with EMF detectors, and I've gone over at length the ins and outs of video and audio equipment; but as I mentioned earlier, some of these tools are of a more arcane nature and we'll focus on that this time around.

The use of dowsing rods for various functions goes back thousands of years. They have been used to find water in new settlements, material objects, fortune telling, and various religious applications. Essentially, a pair of L-shaped metal rods made of brass or lightweight metal are held loosely in each hand and will remain straight or static during normal conditions but, when in the presence of paranormal activity, they will begin to move erratically or cross when directly over, near, or in direct contact with paranormal activity. Interestingly, U.S. Marines even used dowsing to locate weapons and tunnels during the Vietnam War.

Traditionally, the divining rod was a Y-shaped branch from a tree or bush. Different cultures preferred the branches come from particular trees–hazel twigs in Europe and witch-hazel in the United States; branches from willow or peach trees are also common. Both skeptics and many of dowsing's supporters believe that dowsing apparatus have no special powers, but merely amplify unnoticeable movements of the hands resulting from the expectations of the dowser. This psychological phenomenon is known as the ideomotor effect and boils down to basic mind over matter. Your mind is signaling the muscles in your body to make subtle movements that are unnoticeable to the naked eye. Some supporters agree with this explanation, but insist that the dowser has sensitivity to the environment; other dowsers say their powers are paranormal.

The American Society of Dowsers admits that "the reasons the procedures work are entirely unknown."

Research focusing on possible physical or geophysical explanations for dowsing has been conducted in recent years. For example, Russian geologists have made claims for the abilities of dowsers, which are difficult to account for in terms of the reception of normal sensory cues. Some authors suggest that these abilities may be explained by postulating human sensitivity to small magnetic field gradient changes. One study had even concluded that dowsers "respond" to a 60 Hz electromagnetic field, but this response does not occur if the kidney area or head are shielded.

Whatever the evidence for or against, dowsing will undoubtedly continue to be used in the course of investigations. Those who swear by their results will present evidence to support their claims, and skeptics will chuckle at the "superstitions of ages past."

Another example of this type of tool is the pendulum. A pendulum is a small dowsing tool composed of a dangling crystal or metal plumb which

is used to answer questions or find things through psychic energies. Answers are determined by the direction of movement to preset variables; the most common formation is back and forth for yes, circular for no. Pendulums are used in much the same way as dowsing rods and similar to function and result. Due to its design of both answering specific questions and ability to detect or be affected by paranormal activity, the pendulum can be considered a hybrid between the centuries-old dowsing rod and the Ouija Board of Spiritualist fame. Skeptics also point out the high probability of the ideomotor effect in their use.

One device I have to mention, as it's come up in conversation a lot lately, is called the Ovilus.

This odd gadget blends the psychic and the scientific into an all-in-one tool: an EMF, audio recorder, dowsing rod, and K-II that turns EMF into phonetic speech by translating the readings into numbers, and those numbers into words, sounding them out using text-to-speech algorithms via a vocabulary of 512 words. Its various modes include: speech mode, using the environment to pick the words to say; phonetic mode, using the environment to create words phonetically; commutation mode, using speech mode and phonetic mode together, EMF Mode; yes/no mode, to ask questions and get yes or no answers (a digital Ouija?); level mode, to watch the energy change in the environment; and dowsing mode, to work like a pair of dowsing rods. It is equipped with something called the Paranormal Puck. The Puck is designed to aid in paranormal research and meant to be the "center" of investigation. A place to gather, log, track, and maintain the data it watermarks to prevent tampering.

Every time I try to justify this thing, all I can picture is Dug and the other dogs from Disney's *Up*!

Users note that it can be "randomly repetitious" at times by stating selected words for every question asked and every environment investigated.

The first question that comes to mind is how can the inventor of this device possibly test the results? Whatever formula they use to equate EM energy with words would have to start as an arbitrary guess. It would then need to be tested repeatedly to verify the results.

In the end, the most important thing to take away from this is that whatever tools or techniques you or your group are using, as long as it is used correctly and truthfully then happy hunting.

Put away your proton packs and your PKE Valiance devices. While the theories behind Egon's famous PKE meter are based on actual parapsychological research, the toys of Hollywood's *Ghostbusters* are just that. Out in the real world, however, their influence has been ingeniously innovative, often practical, and sometimes just silly.

Investigative technologies in the field of paranormal research have been adapted, tweaked, modified, and reimagined over the past 100 years to further the quest for that paradigm-shifting piece of evidence that will duct tape the mouths of the skeptics shut.

These tools of the trade have evolved from the days of bulky cassette recorders and 35 mm cameras–which were used primarily to merely document data and present it to the world as valid evidence at face value–to include the latest in Bluetooth and other digital instrumentation.

The various tools of today have rightfully expanded their roles to three core purposes: documentation, communication, and authentication.
Advances in digital photography, for example, have reached a point where it has far surpassed the 35mm film of days gone by. The data encoding that is involved serves as a digital blueprint for authentication because it tracks date, time, geo location, and changes or manipulation of the raw images.

Improvements in camera technology are also well noted in the Looxcie LX2 camcorder, an ear-worn camcorder that utilizes Bluetooth technology to pair with a smartphone or tablet.

As a Point of View (POV) camera, it is simply genius in its design, with wireless capabilities that can be viewed from 15-20 feet away. In addition to freeing up the hands for the use of other equipment, the "always on, always ready" feature of wearing it, rather than carrying it, removes the problem of having a personal experience and not being quick enough to get your camera ready to capture the activity.

The newest season of SyFy's *Ghost Hunters* introduced the GoPro2, a camera converted to film in the IR range. Also, the ability of the new "full spectrum" cameras that capture visible light, ultraviolet, and infrared spectrums are just now reaching a level that will blow the lid off of the amount of potential evidence to be found.

The TAPS team also used an app for Android-based devices—a multimeter based on the "Mel Meter" hardware. The interface features an EMF field sensor, ambient temperature sensor, audio recording capability, an LED torch flashlight, and a camera function all at the fingertips. When the EMF field is interrupted, or the ambient temperature rises or falls, an alarm sounds. Pretty neat; and the all-in-one design cuts down on the number of items you have to carry and switch between.

The app uses an Android's built in magnetic field reader to accurately measure surroundings. Not all Android phones or devices may have a built-in Ambient Temperature Sensor, but it is great for

monitoring the overall room temperature while searching for cold/hot spots.

These are all great in theory, but somehow I doubt that the EMF and ATS capabilities will be anywhere near as accurate as those on a dedicated device. Unless you're able to connect an external microphone, most smartphones are impractical for quality audio recording, let alone valid data; and I'll admit that I use my phone quite often to take breathtaking photos while out and about, but anyone who has ever used a camera phone knows that the picture quality in the low-light conditions common to ghost hunting make its use impractical and hit-and-miss–a problem you can't afford to have while on an investigation.

I commend the thought behind it, but the TAPS team dropped the ball by promoting this one. It's just another free geek app that has potential but lacks the hardware behind it to make it scientifically valid. The developer put it best himself, "As with all of my Ghost Hunting apps, please be aware that RESULTS MAY VARY. Neither myself nor anyone else can guarantee contact with ghosts, demons, aliens, or your deceased grandmother."

Here's some of the latest tech that has also found its way into ghost hunting investigations:

Everyone is familiar with the "flashlight test". This was an interesting concept that has now reached laughable status thanks to the folks behind the various shows on television which never seemed to get any results; then lo and behold the boys at TAPS had it happen and now they and every other show out there have at least one incident per week. While this does have the potential to produce motivating results, there are so many variables that can leave even the most zealous investigator sarcastically raising an eyebrow.

Well, someone answered the call and created the Paranormal Touch (PT1) to put a dent in those questionable results.

Place the PT1 on any flat surface, turn it on, and a flashing red light is your phone call to the other side. It uses not one, but two vibration detectors to trigger the unit; and has integrated a Negative Static Detector into the unit to provide additional confirmation of paranormal activity.

Then there's the "paranormal puck," a device that plugs into a computer's USB port and uses "ECM," or "Environmental Communication Mode," technology to "read" energy levels and translate it into text.

Great, now even ghosts are texting. I guess Verizon really does have the world's largest network.

Famed researcher Loyd Auerbach is cynical of a dependence solely on technology. He thinks someone with psychic abilities is far likelier to communicate with the other side than someone with a lot of gadgets around their belt.

Auerbach does, however, feel that technology can provide some clues.

Think about how you might track an invisible boat on a big lake.

"We're really using technology to detect the wake of the boat," he says. "From that, we can infer that some things are going on in the environment, provided we know there's a boat there to begin with."

If there's a common ground that both skeptics and believers can share, it's that ghost hunting is often just as complicated as finding an invisible boat. So we set our sights on the horizon and like the seafaring explorers of old, set off to see what's out there amid the turbulent waters of scientific discovery.

"When Ghost Hunters Go Shopping"
October 2013

Ghost hunting is big business these days. It takes serious cash to attain a collection of data capturing and analysis tools to make your research a success, after all you have to get the latest and greatest gadgets if you're going to troll around at night playing scientist. The trend is such that the owner of a new business in downtown Springfield, Ohio is cashing in on it, confident of long-term success. Darin Hough's Ghost Hunting Source offers shoppers paranormal research equipment in a truly unique niche market.

Perhaps a bit morbidly, Hough explained, "As long as people are dying people are going to have an interest in it." He's not alone. Morticians, cemetery owners, and fraudulent psychics and mediums have been cashing in on the deaths of loved ones for centuries.

The truth of the matter is that there's really no better time to open a store like this. It's possibly the only place outside of the internet where you can buy an electromagnetic field detector, an infrared camera, and a dowsing rod at the same time and take them home the same day.

The questionable "science" of reality television fuels demand for these items as each group and show tries to patent and trademark the latest innovations in paranormal research.

Hough wanted to offer fellow investigators the chance to get their hands on the exact equipment as seen on these shows and play scientist. Sadly, there aren't any proton packs, ecto containment units, or the latest edition of *Tobin's Spirit Guide* available for purchase.

Ghost Hunting Source is in an old industrial site known as the Robertson building and comes with the perfect coincidence–other tenants of the building have told Hough they think it's haunted. And why shouldn't they? Increased foot traffic means their stores will benefit as well.

Naturally, he's hoping to conduct a ghost hunt there as part of the store's grand opening party on Oct. 19 with Miami Valley Paranormal Services–a group which he, conveniently, founded 6 years ago.

Hough promises to instruct newbie ghost hunters on how to properly use what they buy.

"We've got something that's very popular. It's called the spirit box," Hough said, picking up and turning on what is essentially a handheld radio, which he retails between $59 and $79. "The theory is, maybe the spirit could use the white noise to speak through."

Readers of *Across the Great Divide* have heard of these laughable devices before in "*The Harsh Truth about Ghost Boxes*".

Still, this all amounts to big money. Last year alone he claims to have sold more than $220,000 in gear at his online storefront and shipped as far away as Australia and Russia.

Another gadget he's peddling is something labeled E-POD-AMP, a $100 pod that features a series of lights that remain lit until static electricity is detected, and touted as ideal for trying to decipher why the hair on your arm is standing up. Sure, because we all know that static electricity isn't a normal, everyday fact of physical life and only occurs when an apparition is around.

In addition to all of the high-tech gear available at the store, such as full-spectrum video cameras, Hough also carries low-tech, old-world tools including dowsing rods, tarot cards, and sage incense.

Interestingly enough, the one thing conspicuously missing from the shelves at Ghost Hunting Source is a Ouija board, the infamous board made and sold by Hasbro.

"It's a personal thing," Hough said. "I've had nothing but bad experiences with them."

Hough is banking that his store "will turn skeptics into believers."

I would love nothing more than concrete, scientific proof of paranormal activity- which I wholeheartedly believe in the possibility of- but the unfortunate truth is that the greed of capitalism is at its most successful when it prays on the ignorance and gullibility of the uninformed shopper.

There are a number of websites that sell paranormal and ghost hunting equipment and software. These sites offer very little in terms of real-time customer service to answer concerns and questions about the equipment offered and the correct usage of them.

It's a big wheel of supply and demand as shows like *Ghost Lab* act more like hour-long infomercials pushing interest in the newest must-have gadgets. Fans see these cool toys with wide eyes and want to find them at any price so that they can play like the big boys on television. This is why these programs only show those incidents wherein the devices "prove" activity and not the other 99% of the time when it acts erratically or not at all.

Even the money machine known as SyFy's *Ghost Hunters* has jumped into the game. They highlight innovative and glitzy gadgets for an hour and then offer fans the opportunity to purchase their very own at the "Ghost Gear Store", an online shopping site run by NBC Universal. My, how convenient.

Be smart and research these items fully before spending your hard-earned cash. Read reviews, tips, and usage techniques. Question the "science" behind their development and look for analytical data on their results. If you're willing to open your wallet, be willing to open your mind as well. Before you walk into, or web surf, these stores keep the old adage in mind: "A fool and his money are soon parted."

Trust me, their checkout lanes are counting on it.

"Parapsychology's Database Debacle"
March 2014

It's well-known that paranormal research isn't taken seriously by mainstream scientists—after all, there's a big difference between measuring phenomenon like earthquakes and hurricanes—an affect felt and witnessed by hundreds of thousands—as opposed to telekinesis and ghosts sightings, which are often the subjective experiences of an individual.

Not only is there a measurement issue, but there is a records issue.

Advancement in medical research and development, for example, hinges on one crucial component—the existence of a database of verified knowledge and investigative research that is shared, and contributed to, by doctors and laboratories around the world. Without that collaboration, medical research does not progress. A doctor can look up the symptoms of his patient and find that a physician on the other side of the world had a patient with similar symptoms; they compare notes and, at some point, not only is a condition defined, but also a course of action determined.

This is at the very core of the issues facing parapsychology that I have covered lately. Gone are many of the world's leading, official, labs and academic programs—at least in North America; and the few remaining, respected, and professional names in the field are seldom heard from or are departed. Meanwhile, amateur and semi-professional ghost hunting groups are concerned more with competing for exposure and fame, and not with the advancement of science.

This leaves the field with few professional organizations, no official research guidelines, and no reliable, secure central database to pool information that is collected from investigations.

Even if there were such a system, something that I've been an outspoken proponent of for years, there must be a safeguard to certify that the data shared is not falsified, misrepresented, or incompetently interpreted. There have to be similar safeguards for those who are contributing that data. If a chain of people experiment based on fraudulent information, it does a disservice to all and makes the findings worthless. That's a heavy price for someone's time wasted and further ridicule of the field.

There must be an independent group of qualified researchers tasked with keeping contributors to strict submission guidelines and testing and reviewing data to verify the results put forth for others.

These are factors which ruined database initiatives in the past and why any Joe Schmo with a night vision camera and voice recorder can call himself a ghost hunter and get a television show to flaunt his "evidence".

For that evidence to be proven or disproven, and be taken seriously, it must be willingly and freely shared. There are a number of groups out there that refuse to do this.

I contacted a famous restaurant in Detroit that has been reportedly haunted for decades about doing an investigation only to discover that they have an exclusive contract with another group. No other group or research team is permitted to investigate, collect data, or post evidence of phenomena experienced at the restaurant. This contracted group even holds for-profit "tours" on occasion for mutual benefit of the establishment and "credibility" of their own group. The restaurant bilks patrons on the haunting legends and the group gets street cred for it. It's a perfect win-win situation. No one is allowed to verify or refute the group's findings and no one can recreate the exact conditions present when the data was collected to rule out or confirm factors. Not only is this bad science, it's damn insulting.

There are a few ostensible databases on the internet that claim to collect information for scientific integrity, but beware because many of these are hackneyed and trite websites that merely collect folklore and personal anecdotes from often-anonymous responders looking to merely have their stories heard. It's more fan fiction than fact.

A quick search found a number of hits. Paranormaldatabase.com, for instance, seems like a legitimate attempt at such a database but much of the language in their legal disclaimer is highly suspicious and many of the highlighted phenomena have nothing to do with parapsychology or related theories.

Likewise, if a "database" is nothing more than a Facebook page without links to an official external website, or uses gimmicky names or acronyms such as PANICd (Paranormal Database and Research Information), then odds are it's run by amateurs, so beware what kind of information you share with them.

A promising one called ParaDB, created by a Seattle ghost hunting group, is a web-based PHP/MySQL application designed for use by ghost hunting and paranormal research organizations. Its format and design is akin to many mainstream academic and medical forums.

The most serious and legitimate organizations are the American Society for Psychical Research and the famous Rhine Research Center– considered the last bastions of authoritative and academic paranormal research in North America. They publish *The Journal of the ASPR* and the *Journal of Parapsychology*, respectively, and both are world-renowned for the quality of their scientific content including research reports, theoretical discussions, book reviews, correspondence, and abstracts of university and laboratory research papers. I have been a subscriber to both and they hold a special place in my office library.

Until such a time that a verifiable, comprehensive, and worldwide database exists, the ASPR and Parapsychology Association journals will have to carry the weight of scientific discovery, but at least it's a start.

Experimenting with Electronic Voice Phenomena
January 2010

Electronic Voice Phenomena, or EVP, is the observation of disembodied voices or sounds captured on audio equipment that were not present or heard by the observer at the time of recording and are attributed to communication with spirits; this is also related to Electronic *Video* Phenomena. The concept of EVP has had an impact on popular culture as its popularity in entertainment, in ghost hunting, and as a means of dealing with grief has influenced literature, radio, film, and television.

Various explanations have been put forward for EVP by those who believe it to be an example of paranormal phenomenon. These include discarnate entities, such as spirits, communicating on recording media; living humans imprinting thoughts directly on an electronic medium through psychokinesis; nature energies; or beings from other dimensions.

Mainstream science has generally ignored EVP, finding them less than credible, and sites a percentage of recordings that turn out to be hoaxes created by frauds or pranksters. Many also regard the examples put forward by proponents as simply misinterpretations of natural phenomena. These explanations include a variety of known psychological and physical phenomena. The tendency of the human brain to recognize patterns in random stimuli and radio interference are respective examples.

Many Spiritualists believe that communication with the dead is a scientifically proven fact, and experiment with a variety of techniques for spirit communication which they believe provides evidence of the continuation of life. According to the National Spiritualist Association of Churches, "An important modern-day development in mediumship is spirit communications via an electronic device". An informal survey by the organization's Department of Phenomenal Evidence cites that a third of churches conduct sessions in which participants seek to communicate with spirit entities using EVP.

The origins of the formal study of EVP dates back to a period between the 1840s and the 1920s with the Spiritualist religious movement. New technologies of the era, including photography, were employed in an effort to demonstrate contact with the spirit world. So popular were such ideas that Thomas Edison was asked in an interview with *Scientific American* to comment on the possibility of using his inventions to communicate with spirits. He replied that if the spirits were only capable of subtle influences, a sensitive recording device would provide a better chance of spirit communication than the table tipping and Ouija boards mediums employed at the time. However, there is no indication that Edison ever designed or constructed a device for such a purpose.

As sound recording became widespread, despite the decline of Spiritualism through the latter part of the 20th century, attempts to use portable recording devices and modern digital technologies to validate life after death continued to be promoted in popular culture and by a handful of dedicated believers.

Some EVP enthusiasts describe the hearing of words in EVP samples as an ability, much like learning a new language. There is some truth to that. As my own experience with analyzing audio has increased, so has my ability to differentiate between true voices and just simple noise.

Instrumental transcommunication (ITC) is a more general paranormal term than EVP and refers to communication between spirits or other discarnate entities and the living, through any sort of electronic device such as tape recorders, fax machines, television sets, or computers. ITC includes visual and other anomalies, rather than only auditory effects. The term was coined by physics professor Ernst Senkowski, a faculty member of Engineering Department at the University of Mainz, Germany. Instrumental transcommunication has gained no notability within the scientific community and is not accepted within science.

The modern research model for EVP dates back to 1976 with paranormal researcher Sarah Estep. In 1982, she founded the American Association of Electronic Voice Phenomena (AAEVP) in Maryland, a nonprofit organization with the purpose of increasing awareness of EVP, and of teaching standardized methods for capturing it. Estep says she has made hundreds of recorded messages from deceased friends and relatives; Konstantin Raudive; Beethoven; and to other individuals including a lamplighter from 18th century Philadelphia, Pennsylvania, and extraterrestrials who she speculated originated from other planets or dimensions.

In 1997, Imants Barušs, of the Department of Psychology at the University of Western Ontario, conducted a series of experiments using the methods of EVP investigator Konstantin Raudive as a guide. A radio was tuned to an empty frequency and over 81 sessions more than 60 hours of recordings were collected. During recordings, a person either sat in silence or attempted to make verbal contact with potential sources of EVP. Barušs stated that he did record several events that sounded like voices, but they were too few and too random to represent viable data, and too open to interpretation to be described definitively as EVP. He concluded: "While we did replicate EVP in the weak sense of finding voices on audio tapes, none of the phenomena found in our study was clearly anomalous, let alone attributable to discarnate beings. Hence we have failed to replicate EVP in the strong sense." The findings were published in the *Journal of Scientific Exploration* in 2001, and included a literature review.

In 2005 the *Journal of the Society for Psychical Research* published a report by paranormal investigator Alexander MacRae, who conducted recording sessions using a device of his own design that generated EVP. In an attempt to demonstrate that different individuals would interpret EVP in the recordings the same way, MacRae asked seven people to compare some selections to a list of five phrases he provided, and to choose the best match. MacRae said the results of the listening panels indicated that the selections were of paranormal origin.

Skeptics such as David Federlein, Chris French, Terrence Hines, and Michael Shermer say that EVP are usually recorded by raising the "noise floor", the electrical noise created by all electrical devices, in order to create white noise. When this noise is filtered, it can be made to produce noises which sound like speech. Federlein says that this is no different from using a wah pedal on a guitar, which is a focused sweep filter that moves around the spectrum and creates open vowel sounds. This, according to Federlein, sounds exactly like some EVP. This, in combination with such things as cross modulation of radio stations or faulty ground loops can cause the impression of paranormal voices.

The human brain evolved to recognize patterns, and if a person listens to enough noise the brain will detect words, even when there is no intelligent source for them. Expectation also plays an important part in making people believe they are hearing voices in random noise. It is interesting to note that a common practice in EVP techniques is to have some manner of white noise in the background. Older cassette recorders require the use of white noise generators for the purpose of capturing EVP while the modern digital recorders are capable of producing their own white noise.

Interference, for example, is seen in certain EVP recordings, especially those recorded on devices which contain RLC circuitry. These cases represent radio signals of voices or other sounds from broadcast sources. Interference from CB radio transmissions and wireless baby monitors, or anomalies generated though cross modulation from other electronic devices, are all documented phenomena. It is even possible for circuits to resonate without any internal power source by means of radio reception. Such interference is commonly observed in home speaker systems which pick up the transmissions of nearby CB or car radios and transmit the sound through the speakers generally scaring the socks off of home owners, especially if they are home alone at the time.

This effect is so common that the FCC has a rule about it called <u>Title 47, Part 15</u>. Just about any electronics device sold radiates unintentional and random emissions, and must be reviewed to comply with Part 15 before it can be advertised or sold in the US market.

Capture errors are anomalies created by the method used to capture audio signals, such as noise generated through the over-amplification of a signal at the point of recording.

Artifacts created during attempts to boost the clarity of an existing recording might explain some EVP. Methods include re-sampling, frequency isolation, and noise reduction or enhancement, which can cause recordings to take on qualities significantly different from those that were present in the original recording.

In many ghost hunting groups, audio software like Nero's Wave Editor is utilized to filter EVP samples in order to "clean up" the sample and aid in clarifying the sounds for analysis and presentation. This software also helps by displaying a graphical representation of the audio sample in which you can observe a visual modulation of the sound source. I have used this application myself when conducting examination of sound recordings from investigations.

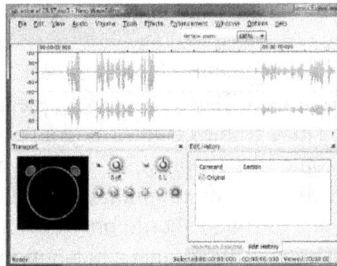

Sometimes the software does help clarify the samples as not belonging to a group member or to shed light on the exact words or phrases captured; other times it has the opposite effect of distorting the sounds. It is for this reason that I keep the original raw data files I collect from investigations unaltered on a separate folder or network drive and work exclusively with a copy. This way if a question arises that the noise reduction or filtering techniques employed are what created the sample, then the original file is available for independent analysis.

Pareidolia and Apophenia are the basis of arguments against the legitimacy of EVP. In the case of EVP it could result in an observer interpreting random noise on an audio recording as being the familiar sound of a human voice. The propensity for an apparent voice heard in white noise recordings to be in a language understood well by those researching it, rather than in an unfamiliar language, has been cited as evidence of this, and a broad class of phenomena referred to by author Joe Banks as Rorschach Audio has been described as a global explanation for all manifestations of EVP, just as I explained earlier

All this history and scientific theory is informative, but the true fun and exploration comes when you can put that knowledge into practice. Below are some common techniques and some tips and tricks of the trade that I've picked up over the years while conducting my own EVP sessions.

In addition to deceased spirits, various paranormal investigators say that EVP could be due to psychic echoes from the past and psychokinesis unconsciously produced by living people. In this respect I would suggest that you go into an EVP session with plenty of rest, a clear and positive frame of mind, and an objective viewpoint. Some people like to 'provoke' or antagonize the spirits into manifesting or communicating. First of all, this is dangerous and should only be done with careful and proper training. This is such a hot-button issue that I will talk about this at a later date in a discussion of its own.

According to parapsychologist Konstantin Raudive, who popularized the idea, EVP are typically brief, usually the length of a word or short phrase. Don't expect complete conversations or lengthy narratives.

Conventional audio cassette or digital voice recorders are used in experiments with EVP. The claim is that ghosts can talk perfectly well but can only be heard on an electronic recording. This means that recording gear has the ability to convert inaudible frequencies into audible ones, which may seem contradictory since they were created specifically for the capture of audible signals within our range of hearing.

Recorders are also useful in recording notes, member movements, report anomalies or mention things for the analyzing team to pay close attention to when reviewing data. The type of tape that is most often recommended is high bias tapes or metal tapes.

You have to use an external microphone when recording. The internal microphone will also be recording the internal gears and motors and this will make your tape worthless. Any sound you hear on the tape could not be used as evidence because of this. Digital recorders may not have any moving parts but still require the use of an external microphone.

It is highly recommended to use an omni-directional microphone (pictured right) that can be purchased at most electronics stores. These provide full-area 360° coverage. Also, remember that the older tape recorders need the addition of white-noise devices; digital recorders create their own white noise.

When recording investigators names it would be wise to have each individual present state their own names, which will make it easier for distinction amongst voices heard on the tape during review.

Skeptics of the paranormal attribute the voice-like aspect of the sounds to the aforementioned apophenia and pareidolia, artifacts due to low-quality equipment, and simple hoaxes. Likewise, some reported EVP can be attributed to radio interference or other well-documented phenomena. This viewpoint is similar to the matrixing effect in still photography, which I'll be covering in the next chapter.

Portable digital voice recorders are currently the technology of choice for EVP investigators. Since these devices are very susceptible to Radio Frequency contamination, EVP enthusiasts sometimes try to record EVP in RF- and sound-screened rooms. Nevertheless, in order to record EVP there has to be noise in the audio circuits of the device used to produce the EVP. For this reason, those who attempt to record EVP often use two recorders that have differing audio circuitry quality and rely on noise heard from the poorer quality instrument to generate EVP.

Each individual will have their own style of gathering evidence. Some groups walk around repeating the same questions monotonously; "Is anyone here?", or "What's your name?", and so forth over and over again...snore...

Personally I like to start with these but then turn to a more conversational style like you would when chatting with a friend or small talk amongst strangers. Start with simple, broad questions, and then push for more personal information–especially if you know who the spirit may be or if you have some knowledge as to the history of the location. It is also important to never whisper. Always keep your voice in a quiet but constant conversational tone and volume. This will help differentiate your voice or those of your fellow investigators from any anomalous source.

When asking questions or when making requests always pause for at least 3 to 5 seconds between statements in order to give an entity time to respond. An EVP is much more discernable when it's not under the rushed sound of your own voice. Have you ever tried to converse with someone who's speaking a mile a minute or not allowing you a chance to respond? You wouldn't like it if it were done to you, so just imagine how frustrating it must be for someone who can't as easily make their discomfort known.

When conducting EVP sessions make note of environmental and astronomical factors. Many of the best EVP recordings and other strong evidence for hauntings occur near streams or in close proximity to groundwater. Early studies suggest that running water may generate a frequency that renders some people more sensitive to psychic phenomena. There is also room for speculation about lunar cycles and increased paranormal activity, or perhaps more awareness of it. With any activity that involves environmental, psychological, or spiritual factors please keep in mind that you do so of your own free will and risk.

As with any research, a lot of it is trial and error. Mix and match, see what works best for you and fits into your comfort level. Also keep in mind that just because a technique has worked well for you, don't immediately discount someone else's technique or be afraid to try something new. Not everyone likes to be talked to in the same way. I like to treat any possible entity as an intelligent and mature adult, and speak to them accordingly; but if it is a child or otherwise, I will try to speak to them on their level. This may provide far better results.

Various examples of EVP can be found online that you can download and listen to. Most reputable ghost hunting groups post their findings on their websites and you are invited to listen and make your own judgment. For some examples that I have caught personally while investigating sites you can go to http://www.deepforestproductions.com/archives.html.

Do You See What I See? Spirit Photography
February 2010

Last month we dealt with capturing audio evidence so this time around it seemed logical to continue by exploring spirit photography. I was honored to chat with Paul Michael Kane in preparation of this article. Paul is a professional photographer and has experience in paranormal investigations. During an enjoyable and humorous chat many topics were discussed.

Many of us have taken a picture from time to time that had strange exposures, lighting effects, or unknown properties and stood there in awe wondering what it was. Paul just about made me fall out of my chair in laughter over what he calls "chimping". This is when someone takes a random picture while out with friends or when part of an actual ghost hunt and stands there mimicking a chimpanzee jumping up and down shouting "OOOH!! OOH!! Look at what I got!!" because they believe that bright splotch on the tiny LCD screen is a ghost when, in fact, it's nothing more than a bug or other explainable source.

Spirit photography is the practice of finding images of paranormal anomalies or spirits on film and is also referred to as psychic photography. It doesn't necessarily have to be night time to get a ghost picture, as many photographs of ghostly anomalies have been taken during broad daylight at various, seemingly innocent locations, proving yet again that ghosts or spirits are around during the day just as much as they are at night.

Light exists within the universe in a spectrum consisting of infrared, ultraviolet, and visible light. Without getting too technical, humans experience the physical world through the visible spectrum but parallel to this are infrared and ultraviolet. Infrared, or "night vision", is electromagnetic radiation with wavelengths greater than visible light and shorter than microwaves; ultraviolet consists of electromagnetic waves with frequencies higher than those that humans identify as the color violet. It is within the boundaries of these wavelengths that spirits make their presence known when they cross over, however briefly, into the visible spectrum and our awaiting eyes.

Just as with EVP, spirit photography has been around since the camera was invented with photographers using many types of film and lenses. For example, Semyon Kirlian discovered in 1939 that when an organic or nonliving object is placed upon a photographic plate and subjected to a high electric current, a glowing "aura" forms around the object and is imprinted on the film. It is more accurate to say that rather than revealing a natural aura, this process produces such. Some effects thought to be paranormal disappeared under more stringent controls, leaving research of Kirlian photography at a dead-end, pun intended. However, fluctuations in the magnetic fields surrounding the subjects can be detected in this way and Kirlian photography has recently come into use as a medical diagnostic device. It also has a popular market at psychic fairs as a sort of high-tech, more expansive version of the mood ring.

The use of digital cameras has become the rage in recent years, in part due to falling prices, ease of use, and the ability to view results immediately on location. Many ghost hunters take hundreds of digital photos at random, using nothing else during their investigation but the camera, and then present the images as absolute proof of the paranormal; by doing so, they make a sham out of the field.

The issues researchers had with digital cameras have dramatically improved in recent years. The problem was with older models and how they operated in low-light conditions. Truth be told, even well-lit day conditions were often grainy, pixilated, and gave questionable results. Referred to as the "orb factor", many ghost hunters would take shot after shot of locations and point to the numerous orbs in the resulting photographs and present it as proof of spiritual activity. These were simply a result of the camera's inability to interpret data correctly, light reflecting off of insects and dust, and areas of the file where the pixels failed to fill in completely leaving blank areas or misinterpreted and warped data. Newer models have all but removed this incompatibility.

Another issue was that a digital image's authenticity would come into question and often pictures were doctored or manipulated in such a way to support a group's claim. To be able to analyze a photo and determine its legitimacy, two things have been needed–a print of the photo and its negative. It has been argued that a digital camera could not provide this and since electronic images could be easily altered, it was impossible to prove they were authentic. To begin with, photographs have been manipulated and staged since the camera was first in use. Manipulation occurred within the camera itself and afterwards during the development process, thus tainting the almighty film negative. Technology has changed and now it is not only possible to authenticate digital images but, depending on the camera, it can be used as the primary photographic instrument in an investigation.

It is now within the financial means of investigators to purchase high-quality point-and-shoot digital cameras that not only offer clean and crisp images that do not have the problems with false orbs but some models also offer night-shot modes. These next generation cameras also offer a way to authenticate the images that is as trustworthy as a negative. One option of a higher-quality camera is access to what are called Raw Data or Meta Data files. These files are uncompressed and unprocessed and an anomalous image that is examined using this option can actually be authenticated with often more detail than in a photographic negative. In addition, the newer cameras also offer access to the information about the images that are photographed. This data is embedded into the image once it is taken containing everything about the camera that took the image including camera settings, date and time the image was taken, if flash was used, the ISO settings, f-stop and aperture settings, and more. If anyone attempts to manipulate the image, the changes are marked as well. In this way, a person trying to analyze a digital image will be able to see if it has been manipulated or not. If anyone attempts to alter the data, it will destroy the image. In this way, it becomes a "digital negative" of every picture that is taken.

Video cameras are another important instrument for any investigation. Unlike still cameras, they provide constant visual and audio surveillance for review and observation.

Due to most investigations being conducted in absolute darkness, video cameras equipped with infrared capability are a necessity due to normal cameras needing a light source bright enough to capture. Many of today's handheld digital camcorders can be fitted with an optional infrared module for night vision recording and are fairly cheap via aftermarket outlets.

With video, any phenomena occurring can be documented in its entirety. This will show the length of time the phenomena occurs, what is happening, the conditions surrounding the phenomena, and possibly even the cause of the phenomena; also widely used are infrared wireless security cameras for the unattended stationary recording of various locations within an investigation area. Paired with a 4 to 8 channel Digital Video Recorder system this is a must have for any investigation covering a large area.

When working with video it's best to set your camera up on a tripod during recording to keep the image or video steady. Don't drag the camera around trying to get something on film, let the spirits come to it. Set the camera up somewhere and just let it run, then view the footage later on. Set your camera to manual focus and keep it focused on something nearby. If it's on auto focus and something unusual comes into view it will spend a lot of wasted time trying to focus in on it, and it usually won't be able to in time, so you won't see what really went by.

Try using external infrared lights to increase the viewing range. When using a night vision camera, the laser dot will appear as a white glowing point, so don't confuse it for something it's not. These can be fitted to a digital photo camera as well so there will be less need for flashes during an investigation. There are Do-It-Yourself kits, but your best option is to have a professional install it.

The holy grail of paranormal research equipment is the thermal imaging scanner. This device is a fusion of the digital video/still camera and the infrared thermometer that allows users to see and record on video what an IR thermometer detects. A deviation of plus or minus 10 degrees is significant for investigation purposes. Should there be a cold spot or hot spot, this infrared technology makes it possible to see the shape and size of the temperature change. These remain perhaps the most expensive items to date costing anywhere from $1,500 to $10,000–second only to the new full-spectrum cameras which can run up to $30,000.

The results of this device are often glorified by groups like TAPS and GHI, but for every paranormal incident that the device shows, there are several that TAPS is able to debunk because of it.

A recent addition to the camera class is the full-spectrum model. Not only can these highly-expensive cameras capture everything in the visible spectrum, but simultaneously in the ultraviolet and infrared as well; the resulting images are black & white with some color hues.

So what exactly should you look for in a digital camera to do ghost hunting with? Make sure it has a good flash and the ISO settings can range from 100 to 1600. By using cameras that range from 8 to 13 megapixels and taking advantage of all of the options available to us, we can actually gather significant evidence with our digital camera that is comparable to that of a 35 mm camera. The Nikon P6000 is a great camera for ghost hunting purposes.

If the camera has a preprogrammed night-shot setting, which is a very slow shutter speed, you may want to invest in a monopod, which can cost as little as $14, and is easily portable on an investigation. With this type of setting any hand-held shot will come out blurry. Some digital cameras will not take a picture if it thinks the area is too dark as well because they require something for light to bounce off of, like a gravestone. Also, use high capacity media cards– an 8 gigabyte card will have plenty of storage space for investigation purposes.

A major issue that needs to be dealt with in detail when working with photography is the matrixing effect, a general term for the natural tendency of the human mind to interpret sensory input, what is perceived visually, audibly or tactilely, as something familiar or more easily understood and accepted, in effect mentally "filling in the blanks."

I was initially convinced I caught something wild when I took this photo (bottom left). It turned out to be the camera's wrist strap. In the example on the right several of the rain drops looked like faces to me.

So there you are, on location and chimping away at your camera screen and exciting your fellow investigators because something looks paranormal on the LCD but later, when the photo is analyzed in larger detail, it turns out to be the reflection of a butterfly's wings or the way the shadows of trees merged together to look like a figure standing in the field. Of the hundreds of pictures that are taken maybe 1 or 2 percent actually produce worthwhile results. Is this to say that cameras are a waste of time? No. When something of merit is discovered it is usually of great importance to paranormal research. Just be sure of what you have before releasing your findings.

A few groups take a conventional camera and shoot away in the hope that something will appear on the processed film. From time to time distortions and anomalies result during development. It should be noted that all such images are well-known an understood effects of photography and of cameras. They happen every day and have nothing to do with paranormal phenomena. Sometimes what may seem like a ghost or other paranormal occurrence in a photograph is the result of physical factors within the camera itself.

We must also discuss the infamous "orb factor" in more detail. The parapsychological meaning of substantiated orbs are floating circular balls of light with color or brightness seen in areas of high paranormal activity and are believed to contain the soul, personality, and emotions of a deceased person or animal. These may be visible to the naked eye or invisible until caught on film. They may also have streaking tails of ectoplasm or glowing energy; most orbs are widely debunked by paranormal researchers as being evidence of paranormal activity.

True orbs produce their own inner light; so would it not seem likely that a true orb will be seen with the naked eye before being caught on camera? After all, a light bulb doesn't just show itself when a picture of it is taken.

The example on the left, of four 'orbs' surrounding a grave, is particularly interesting. While I know it's most likely dust, the formation around the marker–as if standing guard–was very peculiar.

The two examples below show how some orbs are captured.

The example on the right does seem like it's glowing of its own accord but keep in mind that the closer it is to the flash, the brighter it will be. The example on the left is leaving a tail as it moves upward. Dust will generally move very slowly and in an erratic pattern due to wind factors so the cause of this photo is open for interpretation.

There are several examples that Paul suggested as the cause of the majority orbs. The infrared lights from video cameras or teammates' camcorders could be reflecting in. I once was analyzing some video footage of pulsating lights and it turned out that it was the handheld Mini DV of an investigator and a stationary camera's IR modules reflecting into each other thus causing the phenomenon.

There is a term within professional photography known as Bokeh (bo-kah), meaning "out of focus". Many mists are the result of a portion of the lens being out of focus or the operator's finger is over the lens causing condensation. Also keep your fingers away from the flash and remove the wrist strap.

Sensor dust occurs in DSLR cameras when metal shavings from the detachable lenses stick to the sensor inputs of the camera and cause unwanted effects. Sometimes there can also be issues with the shutter sync.

With great jealousy I listened to stories from Paul when he had an opportunity to document a pictorial of a paranormal investigation into Eastern State Penitentiary. Over the course of the event, and during other investigations, he has gathered some great advice for investigators when documenting photographic evidence of the paranormal.

Pull out that camera from time to time and take a few shots. Don't just take a picture in front of you but also over your shoulder while walking. While looking in one direction, quickly snap a photo in the opposite direction. This method of capturing something on film is usually quite effective. Trust your feelings. If you feel something or someone else does, take a picture. If you think you saw something, take a picture.

Take more than one consecutive picture. Rather than taking a quick shot of a stairwell and moving on to the next room, take two or three quick shots keeping the camera in the same position. Most of your typical point-and-shoot models have a sport or burst mode that will ratchet off three to five frames at once. In this way, if you caught something you can have before-and-after shots to help track its movement.

Go with your instincts, but if you've captured something try to debunk it by recreating it. Have someone stand in the same position to see if it was just a reflection or light effect; often it's just our imagination impacting our perception of events. Try to remove as much human element from your photographs as you can. Be aware of reflective objects in the room; with dust and debris the closer it is to the flash, the brighter it will be in the picture. Also be aware of where other team members are and what they are doing. If they are in the hallway taking a picture and you simultaneously take one in a connecting room the flashes will interfere with each other. Control light sources as much as possible and try putting the camera down regularly and set the timer.

Many common settings on the camera can also improve your results. Read the manual that came with your model thoroughly and know how to adjust the settings. Turn off red eye reduction, set the camera for aperture priority, and most importantly turn off the auto image preview. This feature slows down your picture taking because many cameras do not reset for another shot if the screen is active. This also removes any possibility of chimping (*I love this term*) and if you're too busy staring at the screen all night you'll miss something. Leave the in-depth review of your pictures for the analysis phase.

With photos you'll want to import them into a computer and view them on a large screen. Just like with EVP, you'll want to work with copies only and leave the originals safely stored away. You can also zoom in and out of the image to help clarify objects. Be aware of the matrixing effect and go through each one to look for differences in lighting, shades, and shadows consistent with a vortex or apparition. Using the tools available in programs like Photoshop, increase levels and clear up the image as best you can by adjusting for light, contrast, and color balance. The most important thing is to differentiate between reflections and objects that are emitting their own light. Look at how lights and shadows are affected by the objects in question and their positions three dimensionally. Light bends around objects; it does not hover in mid-air.

Two of my most treasured photographs (below) were taken at Goodrich Cemetery, where a lone marker was

hidden in a back alcove; these were taken in succession mere seconds apart. This is an example of a vortex, known in environmental science as a plasma light. There is no satisfactory explanation as to their origin.

So there you have it, a brief, but concise introduction to spirit photography. As with many aspects of ghost hunting, try different techniques and experiments. Remember to try to recreate or debunk any anomalies you encounter and strive toward truth rather than the exciting. I hope *Across the Great Divide* continues to be informative and enjoyable for you and your comments are well-received and always appreciated. As always, and until next time, keep those cameras rolling and always exploring the great unknown.

For more information on the photography of Paul Michael Kane and his pictorial book about Eastern State Penitentiary, please visit his website at www.paulmichaelkane.com

Investigation Techniques

There is a very thin line between the last section and the next. All of the phenomenon, tools, subjects, and controversies are meaningless without a clear method of experimenting and investigating them. The following attempts to simplify a responsible manner of investigating.

Spiritual Work and Paranormal Investigations
March 2010

With the popularity of ghost hunting shows offered today it is difficult for many to look at ghost hunting as little more than a hobby or fad. Parapsychology is a science comprised of a complex set of theories and terminology that has been around for 100 years. I attempt through this column to show both the skeptical and the believable sides of parapsychology and to teach ghost hunting techniques and tips in a manner that most can understand and enjoy while learning about the subject. Some of these terms and concepts are a bit confusing and hard to swallow, but I have faith in the intellect and integrity of my readers. Ghost hunting is a passion, but it must also be respected.

Spiritual work in paranormal research is subject to intense debate. Its tendency to be dangerous and a matter of faith makes it controversial, and its subjective results often questionable. This is the first of a three-part series exploring this area of research. Part one will touch on some basic concepts and practices of spirit communication that can aid in paranormal investigations. Some topics such as soul rescues, possession, atavism, and exorcisms are particularly complex and hazardous. These will be covered in part two. Part three will conclude the series by looking at dream analysis and spirit contact through dreams and other trance states.

As a species, regardless of culture, we have a tendency to anthropomorphize events. This is the human-centric trend of imposing human perceptions and qualities upon spirits and other-worldly creatures or forces, assuming that all consciousness must be similar to humanity on some basic level. Nevertheless, if it is an explainable, natural event—or the presence of an intelligent and free-willed entity, good or evil, it must be dealt with accordingly. The risks to ghost hunters are not just physical due to any environmental conditions, but are also dangerous to their emotional, psychological, and spiritual well-being as well.

True hauntings have an impact on us in many ways. For instance, we begin to either question our faith or strengthen our beliefs. When we are confronted by things that cannot be fought through logic or physical intervention we are left feeling afraid, vulnerable, and fragile. An old saying is that "the dead cannot hurt you, only the living can". But fear can cut far deeper, and leave a scar that no mere Band-Aid can heal.

Sometimes an area of investigation already has the entity present and active but often we've seen that members of a group may attempt to provoke or agitate the entity into reaction by harsh language or insults. The idea here is that if the entity gets emotionally charged enough it can draw on the energy to manifest itself for purposes of the team to document it.

This practice is foolhardy and dangerous for an adept spiritual worker to try, let alone your average person. Those who spend their lives studying theology, whether via academia or as a practicing member of clergy, follow a very simple rule: Don't call up anything that you can't send back!

Most shamans study for their entire lives and their knowledge of their particular craft or realm is second to none. Their knowledge has been passed down through a highly-selective group for generations and most have known for thousands of years what science is only beginning to understand. Shaman and scientist alike both see the same thing but through different terminology and cultural understanding.

There are many objects associated with religious, spiritual, and psychic work. Varying cultural aspects obviously come into play here but we will focus on some of the more well-known and most-often-used of these terms and objects.

An amulet is a symbol with magickal significance worn as a pendant or ring. Examples of these are religious insignia such as the Cross, the Star of David, and pentacles. A talisman, specifically, is a drawing or inscription that is worn or carried for the purpose of summoning strength, power, protection, or the aid of spirits. Runes are ancient characters inscribed upon a stone or clay tablet, signifying some virtue or property, and used for divination. The most well-known pre-Christian runes are of Norse origin and continue to be used to this day by followers of that region's old religion, known as Asatru. An oracle can be a prophet, seer, and visionary. Moses was an oracle because he received and relayed information from God. An oracle can also be a special device or place which aids in prognostication, such as a crystal ball. The Oracle at Delphi is a famous example from Greek history.

Many Native American and other shamanic cultures make reference to totems. These are animal, plant, or natural objects that serve as a crest of a clan or family and sometimes revered as its founder, ancestor, or guardian. A totem can also be the personal guardian or guide of an individual while they walk through life and provide spiritual centering and counsel. A fetish is a shamanistic tool most commonly found in the form of a figurine, animal part, or a pouch containing items with magickal associations.

There are several spirit beings associated with matters of the afterlife; the first such being is called a psychic construct. It has been theorized—and experimentation has been conducted to support this premise—that through directed psychic energies a responsive spirit-like entity can be created, continuing for a time to exist independently. A doppelganger, German for "double-goer", may be seen as an example of a psychic construct or perhaps originating from some other place, person, or factor.

Elementals are beings in magickal tradition and ceremony which are spirits that govern over the four corners of the Earth and are associated with, or reside within, the four basic elements. They are called sylphs (the east–air), salamanders (the south–fire), undines (the west–water), and gnomes (the north–earth).

Guardians are spirits who return to warn family members of imminent danger. These may be deceased relatives who offer messages or aid during moments of distress to their loved ones. Similarly, a harbinger is a ghost of the future that brings warning of impending events.

An influence is an invisible entity of undetermined nature, affecting the inhabitants of a dwelling. This may initially manifest as an inexplicable feeling of uneasiness, then be followed by more definite signs which reveal a haunting.

There are a number of events and skills through which direct and subjective contact can occur with the spiritual realm. The person through whom a spirit communicates is known as a channeler; this is a modern term for 'medium'.

The etheric body is a layer of the physical body, mimicking its design, but composed entirely of energy. This etheric body is similar in many respects to the aura, an all-encompassing invisible and colorful energy field surrounding the outer bodies of all living things. The aura-world is a reflection of our own sphere of existence, composed of the electromagnetic emanations of physical matter, and most likely influenced by thought and emotion. It is another dimensional plane proceeding from one in which we exist. This is directly tied to the phenomenon of the Out of Body Experience. The current terms for OBEs are astral projection or astral travel. When out-of-body you are said to be traveling on the astral plane–the Earth's body double which vibrates faster than the physical planet and penetrates to its core. While in this realm or dimension you have an astral body which is an ethereal duplicate of the physical body, remaining connected by what is called the "silver cord," and experience things in other locations, time frames, or dimensional planes. The energies involved in this practice build up over time resulting in the need for Aura cleansing, a metaphysical ritual that clears an aura of all negative energies which may attract negative spirits.

Previously referred to as ESP, or "extrasensory perception", clairvoyance is the ability to obtain knowledge based on an unexplainable intuition or vision beyond the normal range of human perception about people, things, places or events before they appear or happen. There are some specific subsets of ability associated with a clairvoyant's knowledge or skill.

Clairaudience is the ability to hear spirit communications or noises. Clairsentience is the ability to feel tactile sensations of a spirit's presence, similar to psychometry in many respects but the difference is in that with clairsentience you are sensing the actual, real-time presence of an entity, whereas in psychometry you are seeing reflections of the past–whether visually or emotionally.

Where clairvoyance offers psychic communication more associated with the five physical senses, an empath is an individual who is particularly sensitive to the psychic and emotional emanations of his or her surroundings.

Some paranormal investigators have attempted automatic writing sessions on various occasions and it's an interesting technique to use when conducting an investigation. This is the process of writing material that does not come from the conscious thoughts of the writer and is believed to be direct spiritual contact. Believers say that the writer's hand forms the statement, with the person being unaware of what will be written. It is usually done by people in a trance state; other times the writer is mentally aware of their surroundings but not of the actions of their writing hand.

John B. Newbrough, a New York dentist, claimed that he wrote the book _Oahspe_ by the process of automatic writing on the newly invented typewriter during 1882.

Rosemary Brown, an English housewife, claimed to compose music automatically. She could play the piano, though not very well, and felt that great composers were writing through her. Elsa Barker published a collection of letters during 1914 that she said that she had produced by automatic writing. She said the letters came from the deceased Judge David Patterson Hatch. Her book was reprinted in 2004 as _Letters from the Afterlife: A Guide to the Other Side_. I, myself, took part in a past life regression many years ago. I began writing with my left hand (I am entirely right handed) and signed the name John, claiming to be a farmer in 1762.

Skeptics such as James Randi note that there is little evidence distinguishing automatic writing claimed to be of supernatural origins from a parlor game that is little more than sparks of creativity in the minds of the participants. The Encyclopedia Britannica article on Spiritualism notes, "one by one, the mediums were discovered to be engaged in fraud, sometimes employing the techniques of stage magicians in their attempts to convince people of their clairvoyant powers."

Skeptics assert it is nothing more than the subconscious of those performing the writing influencing their actions, and that there is no evidence messages are coming from anywhere other than the minds of the person holding the pencil. This is yet another example of the ideomotor effect. A 1998 article in _Psychological Science_ described a series of experiments that were designed to determine if people who believed in the ideomotor effect could be shown that it was not true. The paper indicated that attempts to introduce doubt about the validity of automatic writing did not succeed and noted that "including information about the controversy surrounding facilitated communication did not affect self-efficacy ratings, nor did it affect the number of responses that were produced. In this sense, illusory facilitation appears to be a very robust phenomenon, not unlike illusory correlation, which is not reversed by warning participants about the phenomenon."

Even considering the authorship as really belonging to a spirit, automatic writing does not guarantee the literary merit of the works produced, which can range from good to atrocious depending on many circumstances.

Attempts at spirit communication are done most commonly through meditation or hypnosis. Unlike meditation, which is an act of introspective relaxation done for and by the self, hypnosis is a state of profound mental focus that is self-induced although through an external agent, or "hypnotist", who acts as the catalyst for the subject entering this state. As concerns paranormal investigation, hypnosis is sometimes used as a vehicle for past life regression and memory restoration.

The following meditation can be used to facilitate a spiritual journey and communicate with spirits. I want to stress again that mediation and spiritual work takes a lot of patience, practice, and caution. Use the following at your own will and note that your experiences will vary. There are a great many books for learning meditation techniques such as _Meditation for Starters,_ by Swami Kriyananda. I would suggest you start by doing introspective exercises many, many times before attempting to use these on an investigation to contact spirits.

Sit down and relax. Begin to breathe deeply and slowly. In through the nose, hold it, count to three, and slowly release through the mouth, allowing yourself to relax. Repeat this several times until you are completely relaxed. Once you are relaxed visualize, feel, and sense yourself walking through a beautiful forest. You can hear the birds singing in the trees; you can smell the freshness of the forest; you can see the spotted sunlight filter through the leaves. As you walk you are looking for a special tree, a tree that is so large that its branches reach the sky, a tree whose roots are so big that they seem to grow to the center of the Earth. And now, in the distance, you see this Tree of Life. You approach the tree and realize that there is an opening, a small hollow doorway at the base of the trunk. You enter the tree and see that you can climb down the hollow roots into the realm of spirits. You begin to journey down, down...deep down...into the heart of Earth Mother.

When you arrive in the realm, look around and familiarize yourself with this world. If you meet any animals, plants or other spirit beings, greet them and speak to them. Listen for their answer in return. Their answer might come in words or in actions. Speak to them and notice what they say or do in response. Then ask how you might serve them and all of nature. When you are through, thank any beings you saw or interacted with, and return the way you came.

To see spirits, I have also used a specific herbal mixture by combining aloe, pepper, musk, vervain, and saffron and burning it in a smudge pot. A mixture of sweet grass and tobacco can also be used. Use this on the anniversary of the death of an individual if it is a specific person you wish to contact.

Repeat this incantation to encourage the spirit's presence:

"Spirits (or specific name) hear my plea, come forth and speak to me. Come join me in this place, and show yourself in this sacred space."

Repeat the request three times and then wait quietly for indications of a presence. Signs include the scent of flowers or cologne, a cool wind, movement of curtains, and candles going out or twitching erratically. Check equipment to see if anything is reading activity.

Once you feel sure the spirit is with you, do not make it stay long. Take care of your business, say farewell, and thank them for their assistance.

Whether invoking or communing with spirits, dealing with demons or other malicious entities, or meditating to develop keener psychic skills, keep in mind that you must seek out experts and masters of the crafts, conduct extensive study, practice, and attempt all things in moderation. You are solely responsible for your safety, physically as well as mentally, and ultimately your success.

As always, keep your senses sharp, your minds open, your objectivity alert, and your heart, body, and mind healthy. I hope your ghost hunting and spiritual activities continue to grow and encourage you to seek higher levels of understanding and enlightenment.

Spiritual Work and Paranormal Investigations
(Part 2 of 3) April 2010

There is a lot of misunderstanding when it comes to particular parapsychological concepts such as soul rescues, possessions, and exorcisms. Centuries of religious propaganda and dogma have colored our understanding and acceptance of the phenomena and Hollywood has not helped to shed light on the truth, either. In fact, classic flicks such as *The Exorcist* and *Poltergeist* have added so much sensationalistic garbage to the pop culture viewpoint that objectivity has gone out of the window.

Alright, you and your team have catalogued some irrefutable evidence that a location is haunted. Now what?

When a location has legitimate paranormal activity then the real work begins. You can't just come in, run your investigation, and say, "Yep, your place is haunted," then simply pack up and leave the landowner scratching their head. The nature of the activity has to be dealt with in a manner satisfying the wishes of the landowner but also taking into account the well-being of the entities at the center.

Hauntings aren't all interactive or dualistic in character with Casper the Friendly Ghost on one side and the Headless Horsemen on the other. A full spectrum of personalities, characteristics, and levels of awareness come into play.

Hauntings can be broken down into the sub categories of apparition, imprints, and demonic.

In parapsychological terminology an apparition is defined as the projection or manifestation of a quasi-physical entity. The rarest such entity being a full-body apparition; a collective apparition is one seen by more than one person; and a vapor apparition appears as a misty, white form. Apparitions are generally intelligent and thus not only aware of their presence in the physical world but responsive to the individuals and environment around them.

The origin, nature, and reasons behind apparitions and intelligent hauntings are subject to intense philosophical, cultural, and scientific debate. Are they the trapped souls of those who have come before? Why are they still here after their bodies have ceased to function? How do they exist within, around, or outside of the physical realm? Are they truly the intelligent and interactive continuation of loved ones?

These questions are complex and each worthy of separate discussions in their own right; but this is the very reason parapsychology exists–to explore those answers. Even accepted scientific areas of study, such as genetics and medical healing, have scientific, cultural, ethical, and religious debates and considerations.

Dealing with the effects of intelligent hauntings consequently is extremely dangerous and must be handled with utmost caution. There is a major psychological impact on both the entity and those who are affected by its presence. The emotional state of the entity must be determined and research into the deceased person and the events leading up to and following their death must be conducted.

If they are angry or violent this, first off, makes any sort of help on your part difficult–if possible at all. I actually dealt with an entity that died over 100 years ago who had no idea they were dead. According to him, we were the trespassers in *his* house.

I would suggest that if this is the case then religious authorities familiar to the deceased and experts from psychological or sociological disciplines be consulted for how best to proceed. Often these experts can also counsel the homeowners and your team in how to deal with the impact of the activity. Just like with people with addiction or behavioral issues, if they are in denial or don't want help, then nothing you say or do will improve the situation. More accurately, you'll just make the situation worse.

Residual hauntings are more recently referred to as Imprints (alternatively, Psychic Imprints). These phenomena have no observable awareness to their environment and repeat the same behavior over and over again. A frequent analogy is the environment acts like a video recorder, in essence capturing a moment of time and playing it over in an endless loop.

In physics, physical energy has a clear definition, and can be quantified in established units. Psychic or emotional energy has no universally accepted definition, and can't be measured in the precise dimensions. Specifically, physical energy obeys an inverse square law: double the distance from an object and the energy received is one quarter of what it was before. Yet intuitive or emotional energy can be detected by sensitive individuals in another continent as easily as someone in the same room, and prayer appears to reach across equally vast distances with little reduction in its effect. The energy involved is not obeying the basic laws of physical energy.

The theory behind imprints argues that at times when an event has enough physical or emotional charge behind it, the environment collects and traps that energy in objects and minerals present at the time. That is how an imprinted haunting can be tied to a geological location in some instances, and at other times to an object from the area; for example, a chair in which a person can be seen sitting in, whether it's in the same house or a museum halfway around the world.

The initial incident of this energy release is called the flashpoint. There have been examples that the imprint has slowly degraded over time with each playback as the energy trapped within the area slowly releases and dissipates with each occurrence; not unlike old VHS tapes.

Demonic hauntings are frightening to the individual that experiences it and those who are researching it. If you're not afraid, then you're in the wrong field and ignorant. Naturally, the spirit behind this type of haunting will not be human. In this respect normal psychological and biological considerations are pointless. In many instances of hauntings that fit into this category, it was noted that the paranormal events were relatively small. However, as time progressed, so did the hauntings and various types of paranormal and unexplained activity.

Case studies of demonic hauntings have indicated that these types of spirits tend to attack individuals that are experiencing weaknesses on a psychological, physical, or emotional level.

Some of the most common signs associated with these types of spirits include objects being moved, objects seen levitating and moving in and around the area, electrical disruptions such as blinking lights and appliances turning on and off, animals have long had a keen sense to emotional and physical stimuli and may begin to act in an odd fashion, and feeling as if you have been touched or are being watched. These hauntings are intelligent and work to manipulate their environment. This can be a terrifying experience for the individual tormented by this type of haunting.

This phenomenon should be dealt with only by professionals with extensive training, knowledge, and experience in this type of activity. This is not a debatable matter.

Poltergeist is German for "noisy ghost." This is an extremely rare occurrence wherein random objects are moved and sounds are produced by an unseen force, the sole purpose of which seems to be to draw attention to itself. The phenomenon always involves a specific individual, frequently a child or adolescent, with unvented negative energies. Despite the name, it refers to a psychokinetic manifestation rather than a spirit manifestation. Psychokinesis is psychic phenomena generated by a living person. These energies produce loud, destructive results which mimic other paranormal events. Epicenter refers to the focal point or origin of activity used in relation to poltergeist phenomena through unconscious projection of strong negative energies.

As this is a cause of emotional or psychological distress, a more behavioral approach is called for. Psychic phenomenon and 'ghost' talk has a negative connotation in some cultures and that must be taken into consideration first. Without divulging too much information on the paranormal aspects, seek out behavioral health professionals to explore what issues and stress-inducing factors are affecting the individual in question. Often times as the issues are taken care of, the psychokinetic activity will reduce as well.

Time and again, paranormal research groups and parapsychologists will employ the services or advice of demonologists, mystics, and clerics. As pointed out earlier we are dealing with a realm steeped in myth, mystery, and faith. Just as in life there are good and bad sides to the character of a person, so it is in the afterlife. We give names like "angel" and "demon" to these opposing sides to classify and differentiate their meaning, their intent, and their place.

Well-known groups like TAPS use demonologists when the reported events are of a clearly malicious nature. The demonologist will explain to all parties involved what they may be up against or provide suggestions on what to expect and how to react. Often the demonologist will attempt to cleanse or purge the entity or negativity from the home or location via whatever means are agreed upon by the owner. It is not unheard of for a member of the investigative team to take on the additional role of priest or cleric after, or during, an investigation for the purposes of cleansing the location–so long as that person has the training and ability to take on such a role.

A soul rescue is the act of metaphysical intervention between a living person and a spirit whereby the earthbound energy is advised and instructed on how to leave the earthly plane and begin its afterlife existence. This type of cleansing should never be attempted by a lone individual but rather in pairs. This helps provide a system of checks and balances so that the energies involved do not overwhelm or malevolently take over the living or non-living characters in this spiritual dance.

As I stated before, often an entity may be in denial or in an angry or violent state of being. If that is the case, then the first step is to gain understanding and trust between the entity and rescuer to establish a communication level in which it becomes possible to ask permission to help it pass safely and willingly on its journey.

Next will be to determine the history and cause for the attachment be it emotional, physical, or a matter of unfinished business. In matters of unfinished business, the range of difficulty logically lies in what the issues are. Generally, once those issues are dealt with the entity will no longer have a reason or desire to stay. If it is an attachment to a place or object often the entity needs reassurance that the object in question is taken care of or handled in a manner respectful to its desires. Emotional issues are far more difficult to handle. These are hard enough for those among the living to deal with, imagine how difficult it must be for those trapped between worlds dealing with lost family or loved ones; the anger and pain of murder victims, guilt, desire, the list of possibilities is endless and, therefore, so are the methods and avenues needed to resolve those issues.

I mention working in pairs because of the possibility of those involved causing high stress to each other. This leads me to possessions. This is the act of being physically or mentally controlled by spiritual forces, usually in a negative way. The invasion of the human mind by a spiritual or demonic entity for a span of time influences or entirely submerges the personality of the human host. In contrast, atavism is not the invasion of a separate entity but a reversion of the self to an earlier, ancestral type.

Possessions typically involve a resonating, or correlating, emotional connection between host and spirit. Think of this like a parasite which needs a symbiotic relationship with its host in order to survive.

This is really getting into a level of spiritual work that increases exponentially in danger and intensity. The resulting methods are not cut-and-dry or quick-fix solutions. This is not a matter of simply lighting a candle, spouting off a few words, and life goes on.

An exorcism is the ceremonial removal of invading spiritual/demonic entities from a person or dwelling. This practice is present in virtually every worldly culture. The Jewish and Catholic faiths each have a formal "Rite of Exorcism" to be conducted by the respective Rabbi or Priest. This is also known as banishing or clearing.

Countless books, internet sites, and individuals offer up the information and techniques involved in exorcisms, but be mindful of the source when seeking or relaying information. Once again, it would be downright irresponsible to attempt this activity yourself. I have been studying the paranormal and practicing shamanism for 25 years, and even I would never attempt to handle a possession case by myself–nor do I pretend to know all that is involved in their process. These clerics go through years of specific mental and physical training and undergo extensive study in order to be entrusted with carrying out this high risk spiritual work.

A psychic cleansing is a less ritualized form of exorcism, in which a home or site is purified and malicious influences are banished through prayers, spoken as the petitioner moves through the area. Often the silent practice of smudging will accompany or replace the verbal chants used in a cleansing.

The following is a cleansing ritual for use during an investigation where a cooperative entity is encountered and needs help to pass on. This ceremony uses natural herbs and spices to ward off spirit energies via a smudge pot and distributed throughout an area and over people.

Collect smudge sticks of sweet grass and white sage and 4 white candles.

Light the candles and place them at the 4 corners of the room. Using a smudge pot or other fire safe bowl, light the smudge and while sitting in the center of the room close your eyes. Visualize yourself lying in a green field.

There is nothing around you; you are alone and at peace. Feel the warmth of the sun beating down on you.

Say aloud:

"What is dark fill with light, remove this spirit from this site. It is time for you to leave here; all is well. There is nothing here for you now. Go now, complete your passing, and with our blessing fare well."

Always be firm and deal with negative spirits strongly, but always with respect.

For further reading on possessions and spiritual cleansing I would suggest *Reincarnation, Channeling and Possession: A Parapsychologist's Handbook* by Loyd Auerbach. He is a professor of parapsychology and a prominent field investigator in psychic phenomena. Also try *Spirit Release: A Practical Handbook* by Sue Allen.

Through open and intellectual dialogue we learn and grow. In the meantime, as always, hold to your personal journey and be mindful and respectful of the things that lurk in the shadows seeking understanding.

Everyone dreams. Humans and other animals require sleep, and yes, they all dream. Our mental and physical health is dependent upon how we sleep and how we dream.

A common religious belief is that gods speak to us through dreams. For example, according to Islamic belief, Muhammad talked to the angel Gabriel several times through his dreams. As a result, dreams are very important to followers of Islam and a person's dreams are highly respected and private between the individual and the divine.

Some cultures believe we are nothing more than someone else's dream. Some shamanic cultures even say that what we perceive as the physical world is the dream and reality exists while we sleep. Many of the higher levels of mediation and trance-states of altered consciousness allow us to communicate with our animal spirits, spirit guides, and divine entities.

But what if we can communicate with spirits through our dreams the way we do while awake? Many cultures believe that this is a matter of fact, and stories of this type of phenomenon are not few and far between. We've seen that communicating with spirits on the physical plane can be difficult. Factors such as the observer's level of awareness and the entity's ability to affect the physical world come into play. But what about the world of dreams? If we are all energy; if we can understand and communicate with spirits verbally through the collective unconscious; and the fluidic reality of the dreamscape can lead to premonitions, then would it not make perfect sense that spirits can utilize the thinning of the veil between our conscious and subconscious minds in order to communicate in real time and one to one?

Anthropological studies suggest that a belief in life after death originated from dreams of deceased relatives and friends. The research of Dr. Carla Willis Brandon argues that it is indicative of the survival of consciousness in her article *Death Bed Visions*.

Edgar Cayce once said, "Dreams, visions, and impressions to the entity in the normal sleeping state are the presentations of the experiences necessary for development, if the entity would apply them in the physical life. These may be taken as warnings, as advice, as conditions to be met, conditions to be viewed in a way and manner as lessons, as truths, as they are presented in the various ways and manners." However, Carl Jung held that dreams are "the main source of all of our knowledge about symbolism." This means that the messages you receive from your dreams are expressed symbolically and must be interpreted to find their true meanings.

There are different levels and types of dreams. Many are just random images and thoughts from our subconscious thrown into a big mixing bowl. I like to think of general dream states as the mind's version of defragmenting a hard drive. It sorts all the data that is collected, deletes broken or corrupted data, and frees up space by rearranging things. The random images that flash across our eyes during this process are the basis of the majority of our dreams. That's why they range from the hilarious to the downright bizarre. Lucid dreams are those in which the dreamer is consciously aware that they are dreaming and can interact, change, and direct the course of dream events. What we call "vision dreams" and "spirit communication" dreams cross the boundaries of the physical universe…as we know them. These include the telepathic dream, the shared dream, the clairvoyant dream, the astral dream, the psychokinetic dream, and the precognitive dream. The average person spends approximately ninety minutes per night in a dream state. Some of us can remember all, or many, of our dreams; others have trouble remembering that we dreamt at all.

The subconscious mind is very powerful and when we are asleep we are in an alpha state of being. The body is relaxed, and the conscious mind is at rest and usually peaceful. This allows the boundary between our conscious self and our subconscious mind to be more open, thus allowing the dreams to be experienced. Our friends and family that have passed over can and do come back to visit us in our dreams. These spirits will find a way to communicate with us, whether to just let us know they are alright, or to give us a message. I have several such stories in my own family of these encounters.

Some of us are fortunate enough to have developed our natural abilities to the point that we can see and hear them through psychic awareness; others will sense them around them. Sometimes these spirits will manufacture a scent we associate them with.

I was told a story of a woman who communicated with her father and whenever he was around she would smell Electroshave. Whether or not this is evidence of his presence is subject to debate. Smell is the strongest sense tied to memory. If she smelled that or a similar scent then her subconscious mind would think about her father, thus implanting the suggestion in her dreams. Similarly, she could be thinking of her father and her olfactory sense misfires and tricks her into thinking the scent is in the room. Still, there are many cases where a particular scent correlates to paranormal activity.

While we are sleeping they can enter our state of consciousness to communicate with us. Time and time again we have heard stories of those who have passed on visiting and talking with their loved ones while they were sleeping.

The death bed vision that Dr. Carla Willis Brandon discusses is "an otherworldly experience the dying and their family members encounter just before death. The dying will report visions of angels, deceased loved ones, or religious figures mere moments, hours, days, or even weeks before actual death occurs."

One spine-chilling story I can think of involved my grandfather. This was many years ago in Florida and they didn't have a phone in the house–the closest one was down the street. He had a dream in which his brother came to him and was calling out his name. He awoke very startled from the dream and in the disarray he wrote the time, 3:30 am, on the wall. The next day the neighbor came to say that they received a call that his brother had died during the night at that very time; in another incident, my grandfather came to my uncle in a dream and told him if he didn't stop smoking he would die. My uncle quit cold turkey the next day.

Usually when someone visits you in a dream it will be very vivid, often in living color, and you will have total recall of the dream. It will also appear to be happening to you in real time. You will wake up knowing that you, indeed, had a visit from beyond.

There are several techniques you can use to increase your level of awareness and recall of these types of dreams.

First, test reality. Whenever anything in the waking state looks out of place or unusual, ask yourself whether or not you are dreaming. Habitual practice of this technique will make questioning come naturally and to occur automatically while dreaming. Use dream incentives. Outside stimuli can induce consciousness in the dream but are not strong enough to awaken the dreamer (such as lights, perfumes, and music). Incentives work differently on everyone, but the most powerful incentives are those that are tied to something of importance to the dreamer or the spirit you wish to contact.

Before you go to sleep say a simple prayer of protection and ask the person with whom you wish to communicate to appear in your dreams. Communication can occur through conversation or through images and symbols. Interpret your dreams on a personal level. The plethora of "dream dictionaries" available may be nice, but they'll often have different interpretations for the same symbols. When interpreting the symbology of your dreams take into account cultural and societal references and what different objects and places mean to you through your personal experiences.

And, of course, use a dream log to write down all that you remember from the dreams so that you can go back and decipher the meanings and messages later. If you start to focus your analytical skills on every detail at the same time you're trying to recall it things will get lost in translation. Write it down first *then* go back and explore the meanings.

This could take several days to produce results so don't give up if you don't succeed the first night. Try it for at least four nights then stop for a day or two and try again. If after two weeks you still have no results then take a break and try to communicate with someone else. Perhaps the person you are trying to contact is not able or willing to communicate at that time. Perhaps they are not responding because that particular individual has moved on to their next level of existence elsewhere. Keep positive, stay alert, and be persistent and you will succeed.

We now see that speaking with those who have passed can take many forms and levels of skill. Whether you communicate with the other side through the lens of a camera and a recorder, through the use of clairsentient skills and psychic abilities, performing a soul cleansing, or using the power of the subconscious mind to reach those who have crossed over through dreams, a vast array of options now exist to aid in your paranormal investigations.

May your dreams be peaceful, plentiful, and productive. May you continue to expand the boundaries of your personal and spiritual journey. Until next month...

For further reading, Raymond Moody wrote several best-selling books on life after death. I personally suggest *Reunions: Visionary Encounters with Departed Loved Ones* and *Induced After-Death Communication: A New Therapy for Healing Grief and Trauma.*

The article, "Death Bed Visions," by Dr. Carla Wills-Brandon can be read in its entirety at http://www.pararesearchers.org/Psychic/dbv/dbv.html.

Investigation Procedures (Part 1 of 5)
August 2010

If you've been following my column for some time, by now you should have a pretty good grasp of the history of paranormal research and with the tools, tech, and spiritual workings associated with the field. So now it's time to go out and put all of that into practice. In this series we'll discuss the steps for a successful and professional investigation. There are five major components to an investigation: group design and organization, research procedures, the investigation, the analysis, and follow-up.

First of all, don't be overeager. It's okay to have fun and enjoy the experience but keep your thoughts and emotions in check, and the foolishness to a minimum. You don't want a teammate's recorder to pick you up laughing or goofing off in another room and mistake it for something paranormal. Not only is it disrespectful to the home or business owner, the spirits which may be in the location, your teammates, and the field of parapsychology in general–but you are in someone's home for a serious and scientific study, not in a carnival funhouse for a good time.

Resolve your fears and preconceived notions of the paranormal and look at each investigation with a clean slate. Every case will be unique because individuality is just as much a part of the other side as it is here in the physical world. Don't expect anything or compare every little event to previous investigations. Let things occur and flow naturally because in a calm and natural atmosphere you'll have the best chances for capturing quantifiable evidence.

Just as we would be cautious of the spirits' intentions, we must also be cautious of who we invite along on the hunt. Your teammates must be chosen with the expectation of honesty and integrity. Choose wisely when interviewing members for your organization as, unfortunately, there are many dishonest people that may cause more upset in a client's home than the unwanted spirit. Furthermore, you don't want your group's reputation to hinge on the actions of a fool. The Deep Forest Paranormal Society has a specific application that hopeful candidates must fill out if they want to join our team. This can go a long way toward weeding out potential problems. Every member of DFPS must have one of these and a signed agreement clause on file before they are allowed to participate in any official group activities.

Beyond the paperwork and red tape, members should be extensively trained in safety, technology, and protocol requirements. Establish basic parameters and guidelines for central ghost hunting procedures. Also of great importance is the establishment of a leadership structure or chain of command. Create departments like technology, case management, transportation, research, analysis teams, spirituality, and so forth. The knowledge individuals bring to the group may benefit specific departments or talents. Someone with extensive training and knowledge in photography is best suited on the analysis team where they could help debunk anomalous photos, not doing background research. In addition, you don't need the team to be on investigation and have everyone acting like they're the boss. With everyone playing top dog no one will follow and there will be no cooperation, no format, and no professionalism.

I've said many times that the main issue I have with many so-called "ghost hunting groups" is that they're run like an after-school hobby group with no organization, code of conduct, or guidelines. To make sure everyone is on the same page–literally–it's wise to have some kind of written document explaining command structure, job descriptions and responsibilities, and procedural notes. Everyone in my group is given a 30-page manual that was drafted by my lead investigator and explains all that, including attendance requirements and disciplinary steps.

Again, this is a serious scientific field and a job. Granted it's all done on a volunteer basis, but to not look at it as you would a paying job and to have no real control over the way you operate in someone's home spits in the face of science; and it's no wonder the general scientific community doesn't take us seriously.

Many groups say to start with cemeteries. There are pros and cons to this, but mostly cons. These groups do nothing more than go to cemeteries and abandoned buildings and think they're ghost hunters just because they sneak into them in the dark of night on a regular basis and snap a few pictures and audio recordings. The first obvious con is that most cemeteries are closed at dusk, making you an illegal trespasser if you're stomping around after dark. You are free to roam around cemeteries during normal operating times and do as you wish, but you must still be respectful of those who lie in them as well as the rules of those who run the grounds. It is NEVER okay to go into an abandoned building, whether day or night, without the landowner's permission–and that's the end of that.

In many cases contacting the city or church that owns the cemetery or building and presenting your honest and objective intentions goes a long way toward garnering permission. You should also have a client contract that explains what each party's legal and financial responsibilities are. Often having a clause that releases the building's owner of responsibility due to injury puts their mind at ease.

Secondly, cemeteries by design are in urban areas close to well-traveled roads and residential homes. This can seriously pollute any evidence due to various issues. Even abandoned cemeteries in secluded and neglected locations have environmental and noise pollution levels that are both known and unknown at the time of investigation that could skew your results.

It may be a good idea to look at cemeteries as training grounds. Go there during the day with new members to get them acquainted with your group's procedures and techniques. As we know, ghosts don't just come out at night and you might actually catch something. While investigating a new cemetery during the day I actually captured a legitimate EVP.

This would be a great time to build connections between members and see which people work best with each other. Often the personalities of members will compliment another in the organization making for a solid duo for official investigations. This is also a perfect time to familiarize everyone with all the various tools at your disposal–and try out that new full-spectrum camera you just purchased. (*Sorry, I was day dreaming again...*) Every member should have a fully-trained understanding of all the tools used during an investigation so that everyone can get their hands dirty and join in the hunt.

Every city and country around the world has local legends and folklore about famous residents, traumatic and dramatic events, and haunted buildings. Start with some of the more well-known locations and approach the owners or management in as professional a manner as possible. Dress and act like you would for a job interview with a high-end business firm. You'll want to put your best foot forward. Not only will this help you come across as trustworthy but will gain you respect as well. Positive reviews and word of mouth are the best advertisement you can get, and best of all it's free.

You'll want to have certain safety items such as a first aid kit, plenty of flashlights and batteries, water bottles, 2-way radios, and name badges before you start any investigations. Keep these items at center command and readily accessible. Make sure procedures are in place should anyone get injured while investigating and have local emergency numbers handy. Also, make sure to familiarize yourself with the address and specific location (major crossroads) of your area of interest so that in the unfortunate case that emergency personnel need to be called they can get to you as fast as possible.

Most items like water, first aid, and radios are common sense safety items, but I'd also like to mention in detail the reason for badges.

A seasoned paranormal research group will require all members to wear identification while investigating or representing the group in public–even when just doing research. Not only does this present a more professional image but it helps clients, law enforcement, and others know who is and is not part of the group. Remember that law enforcement has the right to request identification; and trespassing on private property can lead to fines, imprisonment, or worse–I've known of ghost hunting groups getting shot at when trespassing in areas at night!

Set up your rules and procedures how you see fit for the specific needs and goals of your group's activities. Just keep in mind that the more structured and professional you are in your design, the more professional you will come off when investigating–and word of mouth can make or break your success.

Just visiting a place that has reported claims of the paranormal and snapping a few pictures, or recording some audio, doesn't mean you've investigated it thoroughly. Once you've landed a big investigation you'll want to research the location as much as possible. Background and historical research is a big part of any valid investigation of the paranormal, especially one with a long history of reported activity. It involves conducting a lot of pre-investigation interviews of witnesses, tedious historical research, looking up prior media coverage, and contacting other investigation groups who may have been there before.

Take organized and concise notes of the names, places, and events you find in your research. Make sure you also site exactly where you found your information and site your sources.

The historical data of the land itself must be looked into including any geological and environmental factors; the architectural history of any buildings that currently or previously existed on the land needs to be looked into as well as the people who occupied the land.

The internet might be easily accessible and convenient, but despite the impression, you can't find everything online–or even accurate information for that matter. Many historical documents haven't been uploaded online at all. It takes deeper insight and work to find the answers you seek, often resulting in a need to actually *go* somewhere and looking over hardcopy yourself. Look through historic documents and manuscripts such as a photographs, maps, newspaper clippings, and artifacts by visiting local libraries and historical societies; tell them what you're looking for and ask for their help–they'll be more than happy to assist you in your search and they'll likely know exactly where you need to look or who to talk to.

Visit the county or city's Registrar of Deeds office and do a background on the history of the building and the land it resides on. Look up the current landowner under Warranty Deeds and trace the ownership of the land backwards. Take note of the various uses of the land itself as it may not have always had a business or home on it. The names you encounter can also help to cross reference with local news about and events connected with the area like epidemics, murders, and other newsworthy events. Often a spirit isn't attached to the home as much as it is to the land itself. There could have been a natural disaster that wiped out an area; famine; war; the possibilities are endless, so extensive research of the area is crucial. A massacre could also provide the emotional fuel for a haunting.

Look for any mineral deposits or environmental factors which could explain the reported occurrences. Theories suggest that water lines, quartz, and limestone deposits amplify paranormal activity so if a home lies on property rich in certain elements take it into account when examining the paranormal activity. Look into known or reported environmental phenomenon specific to the area and consumer energy depots. Consider factories and plants that could be producing by-products that affect health and mental states under prolonged exposure. I should note that electromagnetic fields are unavoidable in residential areas, and despite the theory surrounding so-called "fear cages", there is no significant scientific data to support the claims that it causes hallucinations or nausea.

Research the previous owners. Often families will write or pass down stories of unexplainable events that occur in their homes. Talk with former owners to see if they had any odd or paranormal events occur during their occupancy. See if there are any consistencies in stories from family to family. Sometimes a cause can be found to debunk or explain claims, or further support the claims.

You'll want detailed information from the client about the types, frequency, and level of paranormal activity. Are the events specific to a day, date, or time? Do they recur at specific anniversaries as is the case with residual hauntings or do they interact at random? Do they occur only in specific rooms or do events occur in every room except a certain one? Do they occur only when certain people are in the home? This could even be the result of a friend or someone else who visits the location. There could be something positive or negative about the person that is causing psychic turbulence or giving unrest to the spirit. Also make it clear from the start what the client's desires are. If a haunting is confirmed are they comfortable with it or do they want it removed? They may have a lot of needs and wants before, during, and long after the investigation. These will all be covered later.

It might be uncomfortable for some people to ask or discuss, but it is also important to find out about the emotional, psychological, and physical status of each resident as well. You don't want to claim a place is haunted if it turns out the daughter was hallucinating due to high fever from the flu or the father has a history of paranoid schizophrenia. If during an investigation of claims of apparitions you find a stash of marijuana or anti-psychotic meds in the bathroom, obviously that makes any paranormal activity suspect.

My own group was excited when I received a call to investigate a residential home. The home owner had the most fantastic claims including apparitions, moving objects, voices–you name it. He was adamant we come over ASAP and investigate. Three of us went over for a pre-investigation interview. The house had no working lights, no clothes or food, and Spartan décor. Chalking it up to eccentrics we continued. In the next few days I had received no fewer than 64 phone calls. He had claimed that 17 other groups had investigated within the past 2 years. I contacted some of these groups and was provided a wealth of information regarding the individual and the case. It turns out he has been diagnosed with bipolar disorder in the past and is under no treatment whatsoever. He continues to harass myself and any group spanning five different states.

This brings up a good point. Never, ever, meet with a potential client alone–especially if it's at a residence. You can never know what a person's motives are. Even groups like TAPS have had these situations. They were investigating a house that turned out to all be a ruse because the woman wanted to seduce Jason Hawes in the basement! Always have someone with you and at least one person off site that knows who you're with and where you are. These things can happen, and there are numerous situations that you'll encounter that just have to be experienced. I try to cover as many as I can through this column.

If you're pressed for time you can always do more research afterward to gain a better understanding and clarification of events or to finish up documentation. You'll have a better and more detailed report to give your client in the end. Do some follow up research after the initial investigation if something comes up: like a name on an EVP, or an interesting bit of info, or an eerie personal experience.

Assuming claims have panned out and your background research was fruitful, next we'll get into the hands-on, real-time investigation procedures to make the most of your research experience.

Perhaps it's apropos that we're on to the actual investigation portion and it's October–ghosts, goblins, The Great Pumpkin, and all that jazz. But, I digress...

Only after some level of background research will you be ready to enter the location and explore. Even still, prior to setting up a mutually agreeable investigation date with the client, put some controls in place for the experiment.

For instance, having the client communicate to their neighbors that there will be strangers parking out front and tiptoeing in the dark with flashlights in their home at an odd hour would be a good idea; police showing up with lights and sirens blaring would definitely slow down an investigation. Your client should find a sitter for any pets because you wouldn't want the family cat being responsible for that odd shadow in the basement, nor would you want him jumping out of a closet and giving a team member a heart attack–no matter how entertaining that would be at the time (*no, I am not speaking from experience on that one*). If at all possible the residents should have a place to stay for the night as well so as to not get in the way. Keeping contact with them at intervals during the night and calling them while wrapping up would be preferable to them contaminating evidence with noises or lights to "prove" their claims– whether done purposefully or through restlessness while they wait out the investigation, the less disruption in the home, the better.

The exploration should be fun, but your ghost hunt should be a combination of intelligent analytical skills, respectful scientific approach, an open mind, and respect for the people and possessions in the location–living or not.

It's embarrassing to wander about not knowing what to do next so set up a guideline for the investigation including team assignments, what equipment to use, and areas to concentrate on. Have a command center with one or two people monitoring computers and DVRs, batteries, miscellaneous equipment, and coordinating with everyone to provide time, weather, and investigation updates. Establish a logical time table including session time limits and set-up/tear-down times. Remember that you are coming into someone's home or business. Hours of operation and the personal lives of residents are a factor in the length and time of any investigation; a full night is preferable but rarely feasible. Other factors include how large of an area to cover divided by the number of investigators on hand. Have efficient and detailed investigation goals tailored to the case at hand; expertise is the key to success.

Check local news, weather websites, and the Ghost Weather Station (if you downloaded it) on the day of an investigation to determine any environmental aspects which could enhance or deter your investigation. Check this information again just prior to lights out and at intervals throughout to monitor any changes, and mark the time of each change so that they can be compared with the other results later.

Make sure all equipment is in proper working order, fully powered, and calibrated prior to arrival at the investigation site. You'll want to tape down or conceal equipment wiring throughout the house and perform an extensive tour of the home's geography for both logistical and safety reasons. Additional insurance in case of accidental damage by you or your team should also be considered.

Of particular consideration while hunting in closed quarters would be noise control of the team. Teammates should wear soft-soled shoes rather than heavy boots or shoes that click or squeak, causing unnecessary noise pollution. Accessories such as jewelry should be removed or tucked away if they have a tendency to produce a jingle. Check clothing before going through the home for audible friction as some fabrics when rubbed together can cause a disruption. No baseball caps, brimmed hats, or reflective buttons should be worn either due to the visual obstruction they may cause. Keep in mind–the less reflective, the better.

Make sure all of your team members have had sufficient food, water, and rest prior to investigating. Low blood sugar, dehydration, or exhaustion can lead to the failure of a ghost hunt. It goes without saying that any member that arrives at the location intoxicated or under the influence of drugs should be escorted home. You could keep a water bottle handy; otherwise there should be no eating or drinking while investigating. You will also want to refrain from using the bathroom as well. All the extra noise from the plumbing could taint an otherwise perfect EVP being captured elsewhere. Do your business before arriving to the site. Do not allow smoking during investigations, regardless if the home owner is a smoker or not, as it can be disruptive in various ways such as coughing, and the clouds can appear as mist giving false positives to photographs or other anomalous readings in visual equipment. Suppose reported activity is a lady of the house emanating a floral perfume scent. How can this be detected if someone is smoking in the room or has been near the area recently?

Do a thorough walkthrough of the location to obtain baseline readings with all your equipment to determine normal energy levels as well as to naturally explain any of the occurrences before going lights out. Once lights are out you will look at anything as being possibly paranormal more so than when in the light. Your autonomic nervous system will be active and the darkness makes you revert to fight-or-flight mode. You are more calm and unbiased when in the light. That bent metal plate or loose water pipe is more apt to be seen now than in total darkness when all you can hear is the bang it makes as it moves.

Take baseline readings with a compass, EMF meter, K2 meter, geomagnetometer, ion detector, or Geiger counter; if any spikes occur see if they can be traced to natural causes such as common household appliances, so while investigating you'll know where these are located and can be ruled out. If during the walkthrough all is quiet and readings teeter between 0 and 1 but during the investigation they bury the needle you may be on to something. Also check yourself at this time to see where your equipment registers. Remember you are carrying any combination of watch, cell phone, radio, camera, meters, video equipment, and various metals or gems which can amplify the EMF. You, yourself, are a walking energy field. Perhaps all that equipment is being detected in the next room by another team as an anomalous reading.

Use the equipment at your disposal together or in infinite combinations throughout an investigation. While you track a suspected EMF spike, also run an EVP session asking questions and making demands for specific reactions to determine intelligence. Have your teammate take pictures as you follow the readings around the room. If separate tools record results within the same time index they will support the claim that paranormal activity exists.

Whatever equipment you are using, make sure you are using it correctly. All man-made equipment emits an alternating electrical current. This is what the EMF detects. Remember that a single-axis meter must be turned and tilted along all three dimensional axes to gain a true reading. Also, determine how your particular meter measures activity and analyze the results accordingly.

Some meters measure the amplitude or strength of the field while others detect changes over periods of time.

So if you happen to register a response that is outside the baseline you took prior, stop moving. First, see if it remains stationary or if the point of activity moves around in a random or seemingly-intelligent manner. Try to determine if there is a pattern to the movement that might have a natural explanation, or if the signal grows stronger or leads you to a viable cause. Even though lights are out there is still electricity running through the building powering everyday items from alarm clocks to refrigerators. *All* of these items will generate a response on the meter. If the high readings can be traced to these items, then there is a fault in the appliance that is the cause and not a ghost playing with you.

The K-II meter is a great tool to use in conjunction with EVP sessions because of the question-and-response means of its use. Place the meter on a stationary surface and invite spirits to interact by manipulating the LEDs in a particular manner. You might first notice all lights activate indicating something in its proximity. Set up rules for 2-way communication such as flash once for yes, twice for no, and ask questions accordingly. Having a video camera focused on the meter is invaluable so that a record of the experience can be kept without wasting time going back and forth with writing or voice-recording the responses.

Go with your instincts, but if you've captured something try to debunk it by recreating it. Have someone stand in the same position to see if it was just a reflection or light effect. Often it's just our imagination impacting our perception of events.

I've discussed EVP experiments and Spirit Photography at length previously, so to save time and space, please refer to these links: *Experimenting with EVP*, *Spirit Photography*.

The use of infrared motion detectors can best be used in conjunction with stationary video cameras. If you have a camera set to continuously monitor a particular room or area, set up the motion sensor in such a way as to alert the group to the presence of activity or to possible contamination of the scene by someone or something. If there is paranormal activity present on the footage but the sensors have clearly not been tripped and you can disprove the interaction of a person or animal to the scene you stand a better chance of having valid paranormal evidence. Again, this is why time-stamping every reading and an accurate account of all people in the location is so important to a valid claim.

If the presence of a hot or cold spot is felt, pull out your thermal equipment and begin taking readings. Remember that an IR thermometer can only read something with a visible surface and NOT the ambient temperature of the room. Use the IR thermometer to monitor the drop or rise in degrees or trace the cold spot to a draft or some other logical and natural explanation. For more accurate ambient temperature readings use a thermocouple-based temperature measurement device which can take rapid readings that are essential when dealing with an actual paranormally-induced cold spot.

The Thermal Imaging Scanner can put into visual form what the IR device detects. This not only measures the temperature but shows the varied temperatures of all objects in its aperture. If a teammate is sitting on a couch and the scanner starts reading a heat rise in the seat next to them without some kind of natural explanation you may have evidence of a spirit attempting to manifest itself.

You're now several hours into the investigation and all teams have had sufficient time to explore each area of interest using all the tools at your disposal. It's now time to wrap it up and turn the lights back on. Gather all the teams at command central for a quick debriefing and begin the process of gathering all your equipment. Take special care to gather everything in a logical and efficient manner-don't just start pulling up wires at random and throwing them all in a box. Pull all memory cards out of cameras and place them in special containers along with digital recorders (since it is not possible to backup these items on scene) and correctly save and store all recordings and readings captured on computers and DVR systems. Simply pulling the plug before saving all your data could prove disastrous when you go to analyze it. Once it's gone, it's gone. Night wasted.

Collect any twist, zip ties, or cords used to secure equipment. Make sure that if you taped anything down you leave no residue from the tape and everything in the location is in the same condition and place it was when you first arrived.

After a good rest begins the lengthy process of analyzing the data collected. Don't jump to analysis right away; get some sleep and come back to it the next day refreshed but with the night still clear in your memory.

Any paranormal investigation is only as good as the data collected and how well it is analyzed. Some groups do a lot of their analysis on location as it happens, such as Discovery Channel's *Ghost Lab*. This has a lot of advantages in that you can quickly determine a course of action, or an area of higher priority. You can also readily debunk claims by being able to recreate activity while still on the site. If you get a shadow figure you can quickly refute or substantiate it by recreating it in the very location at the same time and with the exact same environmental factors still in play. The down side to this is that if you're spending all your time analyzing the data as you collect it, you're lengthening the investigation cycles and the potential for missing activity increases. Find a happy medium ground and maybe mix it up with a little on-the-fly and a little post-investigation analysis. Do what works best for you and your team.

Personally, I think it's advantageous to wait until after the group has had a good amount of rest but do NOT put off analysis for a few days. After a good rest, assemble your analysis team and go over things as soon as you all wake up and eat immediately following investigation. When you are refreshed, alert, and your tummies are full, you are able to scrutinize what is in front of you while still having the experiences very fresh in your memory and all team members are present for questions and clarification.

Split the data up between several group members and go through everything together, that way if something anomalous turns up it can be immediately reviewed by the others as to its validity and provides a sounding board for bouncing opinions and reactions off of each other.

It helps to have members who have technical knowledge or skill in key areas. I have knowledge and training in psychology, and can lend opinions related to the psychology of perception; another member of ours is a forensic profiler and a skilled professional photographer. These are just some examples of the kinds of members who can be an asset to your investigations. If you don't have such membership then you may want to present your findings to reputable experts before presenting your final report to your client. Follow up with professional video and filmmakers, photographers, physicists, geologists, and psychologists, for example. To be honest, even if your crack team does have superior knowledge and skills, it's always wise to send the data off to third-party experts for an unbiased second opinion.

Save all raw data files- be it audio, video, or photo- in a secure backup location! I can't stress this enough, as I know first-hand the disappointment of losing valuable data. Put all raw, original data files on a large server and work with copies on individual computers. This ensures that the original is safe in case your conclusions are questioned. Remember to manipulate *copies only* when analyzing them and save anomalous findings in a separate folder.

With photos you'll want to import them into a computer and view them on a large screen. Be aware of the matrixing effect and go through each one to look for differences in lighting, shades, and shadows consistent with a vortex, apparition, or various other paranormal activity. Use the hundreds of tools available in programs like Photoshop to increase levels and clear up the image as best you can by adjusting for light, contrast, and color balance. Be careful not to adjust a picture in such a way that you artificially create the activity you're looking for. Any professional photographer out there with expert-level knowledge of Photoshop can slam your claims. The most important thing is to differentiate between reflections and objects that are emitting their own light. Look at how lights and shadows are affected by the objects in question and their positions three dimensionally. Light bends around objects, it does not hover in midair.

Next are the audio and video recordings. A photo is stable–look away for a moment and it won't change; but, unlike photographs, audio and video must be highly scrutinized and paid attention too; if you are distracted by any means you may miss that fleeting apparition or otherwise ghostly encounter. A glance away from the screen or drowsiness could result in you concluding there was no evidence when one of the field's best verifications is sitting on your hard drive unaccounted for because you were too tired or lazy to see it.

When listening to audio for possible EVPs use noise cancelling headphones that effectively remove other noises from the room you are in. Turn the volume up to a reasonable level as to clearly and accurately listen but not result in a burst ear drum. Too loud is as ineffective as too low.

Effective EVP analysis is something of a special skill that is developed with practice. The more you do it and recognize sounds and effects, the better you'll become, and the more accurate your findings will be. Listen to everything, even long periods where no one is conducting an EVP session. Just because no one in the group is asking questions doesn't mean the spirit isn't asking his own questions or making a statement. Listen for whispers, words, sounds, taps, and bangs. When specific questions are asked listen for intelligent answers. Listen to everything in the context of what is going on at the time of the recording–the conversations between team members, a neighborhood dog barking, cars driving by, or various other factors. Again, this is where time stamping is if importance. If you do have something that is agreed upon by the analysis team to be worthy of further analysis save a copy for later study.

Most digital recorders these days come with bundled software to listen to your audio. If you're lucky it'll be a robust program that has hundreds of tools to bend, pitch, clean up, and adjust the volume of audio samples.

Personally, I use Nero's Wave Editor. The tool kit in this program is simply amazing. With this you can see a visual representation of the file with spikes indicating verified sound. You can adjust for noise reduction, hums, hisses, and clicks. You can also adjust volume. One of the best EVPs I have ever caught might have been missed because the word was said so softly it was almost overlooked. Once the volume was increased it was clearly a direct answer to a specific question! EVPs are among the most spine-chilling, awe-inspiring evidence of paranormal contact.

When looking over video you not only have sound to pay attention to but the visual happenings as well. Just as with photographs, use a large monitor but not so large that you're darting around trying to watch everything. A 15-19 inch screen works just fine. Look for light and shadow effects, objects in the environment being affected by unknown means, strange glitches in the video and defined shapes. That glitch most probably is interference from other equipment but it could also be the manipulation of an entity. This is where the IR motion detectors pay their cost. Suppose the room you're watching has an object move apparently on its own. If the motion sensor guarding the room is not tripped then you truly have something paranormal. Is it a ghost or spirit? Only further analysis will tell, but it is definitely not an easily explainable event.

After you've gone over all the data hopefully you will have something worthwhile to report back to the client with. Even if you don't, it doesn't mean a failure for the investigation. All investigations are a gamble. Sometimes you catch good evidence but most of the time it turns up no valid results whatsoever as far as evidence goes, but each investigation is a learning experience and that has no price tag. Also, the client may be comforted by the fact nothing paranormal was going on and whatever logical, natural explanations you may have for the events in the house could put them at ease. If substantial evidence is found it can also comfort the clients. When friends and family give them a crooked eye they can present professional proof to support their claims.

Quality analysis comes from fair, grounded, unbiased attention to detail and hinges on the knowledge and experience of the collectors and those who go over the results. All of these things come with honest, serious practice. Even the most skilled investigators should still seek second opinions. So don't get too over-excited or discouraged. Be diligent and keep at it. Even I am always learning something new.

We've covered quite a bit of information for carrying out a successful investigation into claims of paranormal activity. You've designed and organized your group, researched the case and its history, carried out the hands-on aspects of the investigation, and meticulously analyzed your data. In this final installment of the Investigation Procedures series we will cover the process of presenting your findings to the client and follow-up protocols you should consider and practice in regards to concluding your investigation.

You and your team have thoroughly gone over every last bit of data not once, but several times to make sure you've covered everything. Every nanosecond of audio/visual evidence, every pixel of every picture has been combed, and all your historical and environmental research has been concluded. Leaving no stone unturned your analysis is complete and you're ready to take your findings back to your client. As soon as possible, within a few days at most, you'll want to return to the client to present your report.

At best you've got some logical answers to debunk some of the claims and with any luck some awesome evidence to present to your client to support their claims. Even if you don't, it doesn't mean a failure for the investigation. Keep a level head and remember that all investigations are a gamble; sometimes you can catch good evidence, but most of the time it results in no valid paranormal evidence whatsoever but each investigation is a learning experience.

It could very well be that a client may be comforted by the fact nothing paranormal was going on and whatever logical, natural explanations you may have for the events in the house could put them at ease. At the same time, if substantial evidence is found it can also aide the clients so when friends and family give them a crooked eye they can present professional proof to support their claims.

First, and most important, thank the client for inviting you into their home or business to investigate. Not only is this professional, but proper etiquette as well. This is a volunteer agreement on both sides, and putting your best foot forward goes a long way toward being taken seriously, being asked back, and having your spotless reputation spread through positive word of mouth. Explain the tools and techniques used in the investigation, procedures followed, and answer any questions they may have about equipment, research methods, and reasoning. If nothing of merit was discovered, explain that this is often the case with most investigations. Explain that it does not mean that the location is not haunted or that their claims have no validity, it simply means the results were inconclusive at that time and the case will remain open for further investigation.

If there were specific results, go through each one at a pace comfortable for both the client and the presenters. Both parties will be anxious and eager to view evidence, but don't rush it. Take time to explain how each piece of evidence was captured, where it was captured, and offer theories to explain the phenomena both natural and supernatural.

It should be stressed *not* to lead with what *you* think a sound or voice could be saying. The power of suggestion can skew the clients' objectivity into seeing things the way you want. Let them hear the recordings first and then discuss what they think it says, or if the sounds are familiar or routine. They know their house better than you do and a paranormal sound to you could be something quite mundane and familiar to them. After they have given their opinion, state the group's position and discuss the reasoning behind it.

After all quantifiable evidence is dealt with then you could move on to any personal experiences that occurred during the night. Point out that these are not "proof" but additions to the lore of the location. If these personal experiences are substantial enough and can be backed up by the evidence then you as a group must determine if you would officially classify the location as haunted before telling the client one way or the other.

Even if you've captured irrefutable evidence of paranormal activity, it should not mark the end of your investigation with this client. Schedule a time and date for future investigations. These could be right away or a few weeks apart.

Many groups visit once for a few hours and that's it. Evidence or no evidence–case closed. This is *not* scientific. Your research methods must be put to the test. Follow-up investigations need to be carried out for many reasons. If no evidence was captured, perhaps the energy wasn't there that night to manifest the results, or in the case of an intelligent haunt they simply didn't want to play along that night. There are many logical reasons as to why nothing of merit turned up, both scientific and supernatural. Don't just assume there isn't any support for the claims just because of one bad night.

By the same token, if valid evidence was found, see if it can be recreated. In the case of anomalous photos, recreate the conditions with those who were present to see if there was some logical explanation that was missed; if you had clear responses to questions or commands on audio, see if they recur. Will they answer the same question or comment again in the same way?

Do one investigation during the day, one at night, and a few using various control situations. After a solid batch of visits, you'll have more evidence to support either theory and will be more confident to make a final conclusion.

When you're ready to close the case, leave the client your contact information and offer follow-up services should circumstances change or warrant third-party assistance. Make yourself available for questions and concerns as well as support services. Suppose after all the painstaking research it is determined that there is no basis for claiming paranormal activity. Let them know you can still be a phone call away to address any concerns they have. If you're final claim is that the place is haunted, many people don't take kindly to spirits making themselves uninvited guests in their homes. If they are troubled by the evidence or the situation, provide spiritual or psychological counseling if you are qualified to do so, or point them in the direction of a qualified person or organization that can. Let them know that at any time if the events continue to occur, worsen, or even dissipate in the wake of the investigation they can call for further assistance. Above all let them know they are not alone and your organization will be there for them now and in the future.

Share your findings with other groups who may want to visit the location themselves and collaborate on the findings. True researchers share their knowledge, not horde it for personal gain. Remember that word of mouth is a powerful thing. Positive or negative comments and reviews could make you or break you.

Keep former or future clients updated and involved on past, current, and future investigations and group activities. In this wired world you'll also want to keep an active online profile to share information about the group, offer services, and post evidence and reviews of past investigations.

So there you are. Your investigation is complete. With the knowledge and experience gained from this outing your next one will be better and your team will strengthen and grow, not only in skill but in recognition.

The "Reality" of Ghost Hunting on TV

In the last few years the rise of paranormal television was equaled by a decline in science and knowledge. I enjoyed the early years when the door flung open and the masses saw what paranormal research involved. Unfortunately, that euphoria was replaced by utter disgust as I, time and again, provided sharp commentary on the "reality" of ghost hunting on television.

Debunking Paranormal TV Shows
June 2010

Flip on your TV on any given day or time and there's bound to be a paranormal reality show on. They're more common these days than the "evidence" they present. The list of current shows includes SyFy's *Ghost Hunters*, *Ghost Hunters International*, *Ghost Hunters Academy*, and *Destination Truth*; ABC Family's *Scariest Places on Earth* (*seriously? ABC FAMILY? *ahem...*); Travel Channel's *Ghost Adventures*; A&E's *Paranormal State*; and Discovery Channel's *Most Haunted*, *A Haunting*, and *Ghost Lab* just to name a few! So, really, how authentic are they? TV is TV. If it doesn't get ratings it goes in the bin.

Television has a long history of playing in the shadows. Remember *In Search of...*? This was a series that explored the strange, weird, and unexplainable. It not only did episodes on parapsychology, but UFOlogy, fringe science, and cryptozoology–which studies things such as Bigfoot and the Loch Ness Monster. It was a well-done show that at times seemed to be an odd mix of PBS' *Nova* and *The National Enquirer*. The show was hosted by Leonard Nimoy, but I suppose Spock had to do something during the 70's. In the mid 90's we had *Unsolved Mysteries* hosted by the incomparable Robert Stack and *Sightings* hosted by Tim White. *Sightings* was by far one of the best shows in television history which dealt with the paranormal. Like *In Search of...* they did stories on parapsychology, cryptozoology, and the like lasting several seasons, in large part due to the hosting talents of Tim

White and the producers who designed the show to be more like a news magazine in the vein of *20/20*. They took great lengths to line up authenticated experts in the subjects they covered with appearances by the likes of Loyd Auerbach and Raymond Moody.

I'm a paranormal investigator. I cannot and *do not* claim to be a parapsychologist or hold any kind of clinical degree in parapsychology. Dr. Phil pretends to be a doctor. I don't. I have years of experience and training but my actual degree is in mainstream psychology. Jason Hawes and Grant Wilson of TAPS may be seasoned investigators who have decades of field experience but they are not formally educated in the theories and statistical analysis of the data they collect. They simply go in, turn on their toys, and interpret the data according to their previous experiences. I *do* give kudos to them for periodically sending data to relevant experts such as film and special effects experts, forensic institutions, and so on.

I use my knowledge of the field and the sciences of psychology and parapsychology and my experience as an investigator to look at all of these subjects objectively. Just because I come in and say a place is or is not haunted does not set it in stone. Let other groups come in and replicate my findings. I may even hold off for a few weeks or months and try again myself.

With that said, just because the folks at TAPS say it is so, doesn't mean it is. They spend several hours on location, gather their evidence, analyze it, and leave. They make no attempt to repeat their "experiments". There is no methodology to their data collection or statistical analysis. There is no empirical replication of phenomenon. No control groups. If they can't duplicate an event over and over again it isn't debunked.

When *Ghost Hunters* first came around I loved it. There were many episodes where nothing happened (which is by far the truth of how it really goes), rarely would they say to a client that their home or business was truly haunted. They would suggest further study is needed to back up the "paranormal activity" they encountered. Now every place they visit is haunted and there is no stutter in their voice when they make the claim. I find most of the show now to be unrealistic and playing to the TV audience only. The thing that's really chapping my ass lately with the folks over at TAPS is the "flashlight test". This is where you take a flashlight with push-button activation and unscrew it just enough to keep the electrical contacts alive but allowing for it to go on or off with the slightest touch. The thinking behind this is if the light can go on and off by command or respond in a logical pattern then it is evidence of paranormal activity. Ok, good theory on paper, but seriously lacking in empirical data or methodology. I've personally fallen prey to this "evidence" while investigating and even posted a short video clip.

Fortunately, I did state that it's not proof, just an anomalous clip. My issues with this test are simple. First of all, there are way too many x-factors that raise questions. No control group and no clearly defined style or size of flashlight. Mini Maglites are the most often used but, pulling from personal experience, even the mini ones have weight imperfections due to their metal casing. What about the flooring? These investigations are in old buildings and homes. Maybe it's off camera and I don't see it, but I never recall any member of TAPS pulling out a level and making sure the surface is perfectly flat and free from drafts or environmental factors. Furthermore, when it first occurred on *GH*, Jason Hawes was quoted as saying he'd been attempting this for nearly 20 years and it finally happened. Public comment was in a frenzy over the event and now this "reality" show has a bona fide flashlight communication every week. Seems to me that the producers are having fun in the cutting room. Jay and Grant have said numerous times in the past that they have no control over how the producers cut the final edit of each episode, thus washing their hands of any deception. Fine, I can respect that- except for the fact that Jason Hawes is now listed as one of the executive producers of *GH*. You can't tell me that position doesn't carry some weight in the decision and editing process.

Then there's the infamous Moss Beach Distillery episode. The TAPS teams heads out to find evidence of the Blue Lady only to discover trick mirrors, recorded voices, and other parlor tricks. I agree that the owners, and the chef, who was to be the on-camera spokesman, should have disclosed the special effects, but I disagree that he is solely to blame. Of *course* Grant found speakers in the walls–it's a restaurant folks! They use the speakers as a PA and to play music in the dining room and other areas. This location has been investigated several times by different teams and experts for decades since the 1930's. Loyd Auerbach, himself, has been investigating the distillery since the early 1990's and many public reports have been made since then that mention quite specifically the "effects" in question. The week before TAPS investigated, Auerbach was called by a producer of GH asking for witnesses' names for the Presidio Officer's Club in San Francisco, as this was the show's main focus this trip. This producer, over the course of conversation, admitted that other points of interest were being considered including the Distillery. Auerbach offered to provide case histories and witness accounts of the experiences there but was cut off; the producer indicated they "had what they needed". Auerbach asked if they were aware of the special effects and he point blank said they were. Television at its finest–deceives its stars, deceives its audience.

Ghost Hunters International seems to approach things not so much on debunking things but instead concentrates on gathering evidence. The tech of the show is impressive. Brain Harnois once disclosed that when SyFy approached them about the spin off they were promised anything they wanted. They got it in spades with the full-spectrum cameras, which have produced some very discussion-worthy results. Their investigation style is much the same as their big papa TAPS, thus falling victim to the things I've already mentioned. They do, however, seem to keep a level head when investigating some of the world's most legendary places and I gained much respect for them when they walked away from the castles of Vlad Dracul and Frankenstein saying "Eh, great place. Not so much evidence." On a side note, I have to wonder just how many languages Barry speaks. Every country they visit he can hold an EVP session in the native tongue.

Paranormal State. As objective as I try to be, I just can't be on this one. What can I possibly say other than this has to be some of the worst "reality" programming out there, let alone a slap in the face of serious investigators and scholars of parapsychology. However, that is just my opinion and I could be wrong. This is at best Blair Witch without the hype and not nearly as entertaining to watch, if only just to cut the crew down at every turn.

I've only been able to sit still for three full episodes of the show, and who knows, it may have gotten better but I swear I saw fishing line moving an object in one case. I've seen clips here and there, laughing out loud to YouTube, much to the utter confusion of fellow Panera Bread patrons. Its crew is all about "feelings" and subjective experiences that can't be seen on camera let alone quantified with scientific instruments. The team is so wishy-washy that I wouldn't count on them to battle a turtle let alone an Elemental. Their "acting" is monotone and boring. A critic for the *Boston Herald* once wrote, "There hasn't been a more suggestible crowd gathered since the last *Crossing Over* taping with alleged psychic John Edward." (*Don't even get me started on him*)

The show has faced intense criticism both from the media and viewers who question whether the "activity" is real or faked. The line nauseatingly staggers between documentary and sitcom and dialogue so ridiculous it must be scripted entertainment.

Way over at the other end of the spectrum is *Ghost Lab*. The level of scientific scrutiny is both respectful and refreshing. This show has some of the most impressive technical equipment and knowledge of its application I have ever seen in paranormal reality television. The team rolls on to location with a mobile analysis RV that would put any military recon mission to shame. It's akin to comparing Grant Wilson's K-II meter to the Deflector Dish on the Enterprise.

What I find refreshing is that week after week you sit there excited to find absolutely zilch at the end of the episode most of the time. A big criticism made of the show is that it tries to be *Ghost Hunters* by going to all the places Jay and crew went before. Again, that's why I like it. A fresh set of eyes with different equipment to duplicate, confirm, or debunk, those who have come before. A new perspective.

I wish these various paranormal teams would collaborate on a central show or other forum where all their evidence can be cross examined and cross referenced. Perhaps then we might shed some light on the truth. Honestly, if *all* groups (televised or not) would combine and share their data for the sake of science instead of their ego and fame it might serve the greater good.

Our television shows and the casts that populate them are like extended family. Viewers have a strong kinship to people on a screen that they've never met, if ever they will. We're fans of one show or another because the people and places that we see each week are comfortable and engaging. Whichever show you like, in the end it's all entertainment. That's why it comes back season after season; if it stops being fun to watch it goes the way of *Heroes*. If it inspires and motivates you to find the answers on your own, then great. Don't just take the word of the investigative team. Look past the Hollywood hype and see the truth laying somewhere between the cracks.

Ghost Hunting and Entertainment
April 2011

I can no longer stomach ghost hunter shows. I've panned them in the past, and they've continued to get as stale as bread with a hole in the bag–slowly drying out to become as brittle and useless as the "evidence" they purport to bring to the academic dinner table.

It's the same boring thing week after week, and show after show: a hapless team goes to a location, sets up some toys, flips off the lights (never mind the fact that they don't cut the power, a possible factor, but just the lights), flips on the clichéd monochrome night vision, and tries to scare a viewing audience into believing in ghosts through theatrics and really bad acting. Gone is the science, to be replaced with a Hollywood sensationalism that malnourishes the brain that is in search of something of more substance.

The incident that did it for me, personally, was the *Ghost Hunters* 2008 Halloween special. I sat there, anxious for hours, awaiting some evidence to cross my seasoned senses. What did I get? Grant Wilson's hood gets tugged! It seemed so awesome at first, but turned out to be debunked by several different people as trickery and wires. I felt cheated out of 6 hours of my time. Never again, I vowed. This "reality show" gets more scripted and fake as time goes on.

There's no rule that academia and entertainment need to be mutually exclusive; in fact, they could learn a lot from each other. But the simple fact is that true paranormal researchers get a bit irritated by the celebrity status of these so-called "experts." It's not that they must have PhDs after their names in the credits, but the sad truth that because they're on "reality TV" the viewing public is falsely led to believe they are experts by the networks.

It's never been easy for those who choose to study psychic phenomena. Mainstream science views them with a deep disbelief in much the same way alchemy was looked at before it became known as chemistry. Those few parapsychologists with PhDs fortunate enough to have a home at an academic body continue to search for irrefutable proof that paranormal phenomena really do exist.

And they've been doing it for a long time. In the late 1800s, Harvard psychologist William James risked his reputation by studying things like "crisis apparitions," a clairvoyant event in which final farewells and messages are claimed to be received from the departed before it is consciously known they are dead. In the early1900s, Joseph Banks Rhine helped to pioneer the study of ESP by founding a once-prestigious parapsychology department at Duke University.

Parapsychologists advise that these ghost hunting shows are doing a gross injustice to those pioneers by intensifying the troubles that have historically plagued the field because the scholars without a show at their disposal are replaced by the glitz and glamour of Hollywood, and the promise of a client's chance to cash in their 15-minutes of fame.

Many parapsychologists rely on a substantial part of their income and research funds coming from speaking engagements and lectures. But the well is running dry.

Loyd Auerbach, a noted author and field investigator with over 25 years of in-field research, is one of those parapsychologists to feel the squeeze from Hollywood. In 2006 he was paid for 14 events in autumn alone; that number dropped to five for all of 2007. The year after that? Two. In 2009–one.

Once a prominent and sought-after man in his field, he's been traded for these celebrities because TV stars bring more ticket sales then the stereotypical scholar with a turtleneck and suit coat. "I was making a good part of my living lecturing and doing events. Now the TV stars are getting all the lectures," he said. "It's been difficult to pay my mortgage."

Over the years of my own pursuit of the unknown I get frequently asked, "Why this fascination with having all the lights off?" Simply because it adds to the creepy factor and draws the viewer in like any other B-grade horror flick. Dr. Andrew Nichols, an expert who did research for the U.S. Army (and who received the only grant ever awarded to study alleged hauntings), believes that these shows also push questionable science on the public. Nichols provides a list of what he calls bad science in these shows: Investigations always take place at night. Why ghosts would only come out then is illogical? How can anyone be a good observer in the dark? Instead, as Nichols puts it, "they just run around like little girls."

"We get painted with the same brush," said John Palmer, PhD of the Rhine Institute, one of the only parapsychology institutes remaining in the country. If there's one thing that skeptics, mainstream scientists, and the general public can all agree on, it's that this image of the bumbling reality TV amateur is the first thing they all think of when the subject of ghost hunting comes up in conversation.

GhostDivas, a popular podcast, interviewed former TAPS member Donna Lacroix back in 2009. During the course of the segment she made some very interesting revelations. *Ghost Hunters*, first and foremost, is completely entertainment and everyone is a backstabber. Jason Hawes and Grant Wilson are "the kings" in front of, and behind, the camera. Long-time fans of the show will remember Brian Harnois. Jay and Grant, she asserts, had their whipping-boy in Brian, and she feels Brian was exploited to the point of mental abuse. Donna gives insight into just how "brutal" and "mentally abusive" they were towards Brian. She even addresses the rumors that the network employs a staging crew. Anyone remember the Moss Beach Distillery fiasco?

You can listen to the full interview yourself at http://www.ghosttheory.com/2009/11/16/donna-lacroix-talks-secrets-behind-ghost-hunters.

Diehard fans of these shows must understand that true parapsychological research is not, and cannot, be done through a weekly reality show where ratings and advertiser revenue are the real decision-makers of the show's survival. It is done through tedious, often boring, study and analysis over a period of time determined by each individual case. To rush through countless hours of data for a final report a day or two later is just bad science.

All of this can be summed up with a classic scene from 1984's megahit *Ghostbusters* when Dean Yeager comes to kick the hapless trio off of Columbia University's campus.

Dean Yeager: The University will no longer continue any funding of any kind for your group's activities.

Peter Venkman: But the kids love us!

Yeager: Dr. Venkman, we believe that the purpose of science is to serve mankind. You, however, seem to regard science as some kind of dodge, or hustle. Your theories are the worst kind of popular tripe, your methods are sloppy and your conclusions are highly questionable. You, Dr. Venkman, are a poor scientist.

Peter: I see.

Yeager: And you have no place in this department or in this university.

I think that's the only time Hollywood and the scientific community will ever see eye to eye on this issue.

Hmm, maybe there is some common truth in this Real vs. Reel conundrum after all.

Businesses featured on paranormal television shows are attempting to milk their 15 minutes of fame long after the cameras have left the building.

I get that–to a degree.

They're in business to make money; but they do so by making a mockery out of the science of paranormal research with deceptive business practices.

These businesses aren't interested in the science behind the paranormal activity–whether it's fact or fake; and they're using shows like *Ghost Hunters* and *Paranormal State* to get advertising for their struggling business, then cashing in on that by playing up the results of those investigations by making themselves into a tourist attraction.

Followers of *Ghost Hunters* remember well the "Moss Beach Distillery" incident. For those that don't, let me bring you up to speed.

The Moss Beach Distillery has been reportedly haunted for decades. Some very well-known and well-respected parapsychology professionals, such as Loyd Auerbach, have conducted methodical scientific investigations there with some interesting results; some of those investigations have been written about in the world's leading parapsychology research journals.

The restaurant made a tourist attraction out of that history; and has since installed gadgets to "enhance visitors' experiences." Okay let's be fair; we all know that you can rarely set paranormal phenomenon by a watch. Patrons will want to come because of the chance of being spooked, but that isn't going to happen to everyone all the time; so they installed speakers in the bathroom and several other parlor tricks in and around the restaurant to both excite and humor their customers.

The members of TAPS had a fit when, in the course of their investigation, they discovered the gadgets. They weren't on, but there is a point to be made that it could have tainted evidence had they been on; in the case of EMF, the mere presence of the speakers could have influenced readings whether on or off.

Here's where the politics of "reality" television come into play.

The members of TAPS made public statements asserting that they never knew the items were installed, but there is documented evidence that indicates that the producers of the show were specifically informed by Loyd Auerbach, and several others, during background checks on the distillery that the gadgets were present. The show's producers never told the members of TAPS about it, assumingly, so as to make compelling television.

There you have it, then. *Ghost Hunters* gets ratings for the SyFy channel and the Moss Beach Distillery gets some national advertising. Everybody wins, right?

Well, that is everybody but the scientific community and the public who were deceived by the incident.

A little over a year ago I was asked if I would be interested in getting a group of friends together to go on a haunted tour and ghost hunt at Waverly Hills. It was going to be a fun and unique memory for someone's 30th birthday. Sure, the idea intrigued me, but there was a catch. To keep the cost down, the tour requires a group of at least 10 people paying $100 each.

So, let me see if I have this straight. Waverly Hills (a dilapidated and abandoned mental asylum) gets the fame of being proclaimed haunted by the pseudo-celebrity folks over at TAPS; now the property owners are cashing in on that notoriety by charging $1,000 to take groups of tourists on a thrill ride by letting them conduct a 5-hour ghost hunt. This is, on its face, nothing more than a glorified and ridiculously expensive, Halloween-style haunted house attraction.

The basic format of haunted tours impacts any "investigation" conducted at such an event.

Let's also be honest. The intent of some tourists won't always be mature or even scientifically-themed; the mental state of those conducting the investigation will impact any results collected. I'd be willing to wager that a group of high school or college students won't have serious science on their mind. EVP sessions will also, undoubtedly, be filled with a lot of gasps and excited shrieks over every gust of wind.

If paranormal phenomena are caused by, or derived from, the energy created at the time of an emotionally-charged event, then having large groups of people constantly imprinting their own energy over that being investigated will affect not only that session but the following ones as well. This is a serious risk in any investigation, no matter where or how small.

The chance for tainted empirical evidence goes up with the size of a tour group in comparison to the size of the building being investigated. If you have ten or more people walking around a building the chance for the noises caused by those members, the flashes from their cameras, and the various effects that their other equipment may provide can seriously raise questions as to the validity of any data collected.

There's also another parallel trend I've been meaning to address. That of some ghost hunting groups having exclusive contracts with reportedly haunted locations.

Here in the Detroit area we have a very posh restaurant known as The Whitney; it is also known for being haunted.

I had contacted their management at one time to request an investigation by my group. I was well received by the person I spoke to, but a short time later I was emailed by another person stating that an investigation would not be possible because The Whitney had an exclusive contract with another Detroit-based paranormal group and that they were the only organization allowed to investigate at that location. This supposedly scientific research group also regularly hosts haunted tours for profit benefiting both the restaurant and the ghost hunters.

Any evidence that this group presents from this point forward would be highly suspect because no one else can come in to refute or confirm the findings.

DFPS is not-for-profit, and will always be so. Science is not for profit. Any group that charges to do an investigation is not a scientific organization; they're money-hungry glory hounds. So does any group that maintains an exclusivity clause with a location. This smacks in the face of one hundred years of diligent research by academics and real scientists. This group isn't in it for the science, they're in it for the fame.

Another local spot is Camp Ticonderoga, an historical home in the city of Troy with a checkered past turned into a restaurant. The location is reportedly haunted by a spirit named Hannah–a fact that is proudly extorted on the restaurant's website. However, good luck landing an investigation there.

I tried, as did a few other ghost hunting groups, to come in and investigate to confirm the findings. My requests were treated nonchalantly until they weren't answered at all.

The restaurant wants to claim paranormal activity, but not allow independent groups in to investigate to confirm that.

This type of business practice is steadily becoming the norm, and it really frosts my cookies. To deny the objectivity and fresh perspective that can come from different research styles is just bad science.

You can't have a monopoly on knowledge. Something seems fishy, and I don't mean the menu.

"Ghost Hunting: A Eulogy"
September 2013

A recent death went largely unreported.

There was no outpouring of grief. No endless video loops in the network news feeds.

It went unnoticed because it was a death that had been slow and arduous, occurring over the last 20 years. Like a bedridden relative whose end was known to all, so that when it finally came it was of no surprise and everyone was glad to see an end to the suffering.

I am talking about ghost hunting.

In the period between 1880 and 1980, paranormal research was a science taken, at least in part, seriously among academics who loved the intellectual debate, albeit with a skeptical chuckle amid the discussions that took place in universities and scientific societies around the world. Hey, even the study of psychology and the treatment of mental illness was considered fringe science for decades.

In the 1990s the reporting of paranormal activity became popularized in serious news-style shows like FOX Network's *Sightings*. It became serious and respected. Then the new millennium dawned and reality television took over the airwaves like a cancer, spreading with reckless abandon and swallowing intelligence and reason within its darkness.

Ghost Hunters started with such promise. The first two seasons saw most places debunked and not much occurred by way of evidence. It was more telling of what really goes on in front of and behind the scenes of ghost hunting groups. Then it became popular. The network suits saw dollar signs and everything changed.

Over the years that show has become less about the science and more about the hype. *Ghost Adventures*, *Ghost Lab*, *Most Haunted*, *Paranormal State*, and their like all followed suit with melodramatic acting, gimmicks, and sensationalistic nonsense.

Science took a back seat to ratings.

Ghost Hunters, for example, has a staging crew that sets up a place before their arrival. They've now added gimmicks like taking a dog with them on investigations. Forget the fact that Jason Hawes repeatedly said in the show's beginning that they respect but stay away from mystical techniques; let's also keep in mind that anyone knowledgeable in psychic abilities knows that cats are the animal to use in a ghost hunt, not a dog. I'm serious on that point.

Their latest Hollywood makeover is to have comedian Jerry Seinfeld do the show's narration and they routinely have "guest investigators" who are actors on other SyFy series as a cheap and easy cross-promotional advertising tool for the network.

"I'm not a scientist. But I play one on TV."

The "fans" just eat it up. Meanwhile the anti-science cancer continues to eat away at their critical thinking skills. The fans emulate their television heroes by breaking into abandoned homes and cemeteries in the dark of night and "investigate," then have the grapefruits to proclaim to the world that they are scientists with unquestionable proof of paranormal events.

All of this led to what happened on August 21, 2013. That's the date that ghost hunting lost its battle with brain cancer and was pronounced dead at 21:00 EST–because that's the moment the season finale of SyFy's *Paranormal Witness* aired.

None other than Sebastian Smith himself directed the episode. Horror fans know who he is. For those who might need a reminder, in 2007 he was one of the minds behind the low-budget UK flick Dead Wood.

Seriously. A horror filmmaker is now directing "reality" television that's supposed to be about science and fact. If there was any lingering doubt that reality television of any type wasn't scripted nonsense it must surely be gone now.

The episode involves the "Lynchville Secret" and tells the story of a single mother whose dream of a new life for her and her daughters is disrupted by the presence of malicious spirits from the Old West and a one-hundred-year-old secret.

Boy, if that doesn't sound more like a summer blockbuster and less like a scientific expedition.

I have books, parapsychology research journals, and other reference material going back decades in my office library. Not one of them has even a mention of the so-called "Lynchville Secret". If this story/activity has been around for a century then surely somebody must have said or written something by now. Therefore, I took to the internet. No good there, either; just endless pages and links with reviews and airdates for the SyFy episode.

Shows like this need to have a "Do Not Resuscitate" clause in their contracts.

These "gimmicks" that these shows have been reduced to using are like feeding tubes and life support, keeping these shows going long after they should have been laid respectfully to rest. When you produce a show that's supposed to be about learning and science and instead it must resort to smoke and mirrors just to keep it relevant, then it's time to just pull the plug. It's the respectful and honorable thing to do.

So it is that we lay to rest the fad of ghost hunting. May it go quietly into the night and slip gracefully over the veil that lay across the Great Divide.

"All Good Things..."
July 2014

I sit outside with our dog, feeling the inviting rays of the Sun upon my back, enjoying the crisp air as I type this; and the feeling turns bittersweet. Gone are the long, dark nights of winter as the stirrings of spring harken warmer and sunnier days and the flora and fauna are returning in a triumphant display of color and vibrancy of life's ever-turning wheel. I suppose that it is fitting that our lives, too, change with the seasons.

What I have accomplished has led me to a wonderful opportunity and I have reached a point where my personal and creative endeavors are at a crossroads. I am pleased to share that I have been granted a role in the upcoming film Aladdin 3477. Additionally, my other writing projects have moved up to the next level and I am both excited and exhausted as I juggle the responsibilities of daily and professional life.

Over the last several months I have contemplated what is to come in the next chapter of my life.

I began this column in the fall of 2009 and every month since I have been honored and blessed to share and discuss with you all a common love and interest in all things paranormal.

I have had many wonderful and thought-provoking discussions on these subjects with fans and friends alike; but, sadly, the time has come for me to say goodbye and put my full attention on other projects that I have put off for far too long.

I want to extend heartfelt gratitude for sharing in this journey as we walked across the Great Divide into the realms of the unknown. Know that I do not come to this decision carelessly or thoughtlessly. I consider myself truly grateful to have been part of the *Pagan Pages* family for these many years and to share my knowledge of and experiences in paranormal research.

I hope you learned from me as much as I have from each of you and take that knowledge with you along your personal, spiritual, and academic paths. Always remember that it is on those rare moments when these three paths cross that true wisdom and awakening occur.

Again, thank you and blessed be.

Wolf

Parapsychology Tomorrow

My daily involvement in paranormal research may be on a hiatus, and I may have become a bit (*greatly?*) disillusioned by what reality television has done to the field; but I am still a firm believer in the possibilities explored here, and in the opportunity for science and knowledge to be dragged kicking and screaming into a new paradigm.

There will be many who still insist that every speck of dust is an orb or entity; just as there are those with intriguing material who will be laughed out of the room for believing in such "nonsense".

Perhaps one day I may grab some equipment and venture once more into the mist in search of the unknown and misunderstood.

In the meantime, I remain indoors being a thinker- because science isn't easy and I have some free time to ponder the mysteries of the infinite multitude of universes.

Speaking of which, there is a theory that I have been batting around with a racket inside my mind for a while now that attempts to use quantum physics to explain paranormal phenomenon and I put it to you, my dear readers, to help put the puzzle together.

Quantum mechanics may actually have proven that paranormal phenomenon is scientifically possible. M-Theory, or membrane theory, attempts to unify every version of string theory in quantum physics under a single heading.

Okay, what the heck is M-Theory?

Perhaps not so surprisingly, there are things that even physicists can't figure out—one of which is the issue of quantum gravity. Our current human understanding of it is still based on Einstein's Theory of Relativity.

Most of us are, fundamentally, familiar with only three dimensions: height, width, and depth; but wait, because in Einstein's Theory, space and time are actually combined in a fourth dimension known as spacetime. As a result, gravity is a logical consequence of the geometry of that spacetime.

Why does this matter, and what does it have to do with ghosts? Well, it makes it more mathematically pliable. Hold on. This is going to be a bumpy ride but it will make sense.

Science, specifically quantum physics, has established some rather interesting things in recent years that parapsychology scholars have theorized for over a century:

- Your consciousness affects the behavior of subatomic particles
- One of those, the Tachyon particle, can move backwards as well as forwards in time and can be in all possible locations simultaneously
- The universe is splitting every Planck-time (E^{43} seconds) into billions of parallel universes
- The universe is connected with faster-than-light transfers of information

For M-Theory to make sense in this discussion, though, it *requires* extra dimensions in order to remain consistent. In fact, there are eleven dimensions we are dealing with.

In string and supergravity theories, a brane is a physical object that generalizes the concept of a point particle to the higher dimensions. For example, a point particle can be viewed as a brane of dimension zero, while a string can be viewed as a brane of dimension one. It is also possible to consider higher-dimensional branes. These branes are dynamical objects which can propagate through spacetime according to the rules of quantum mechanics. They can have mass and an electrical charge. Physicists often study fields analogous to the electromagnetic field which exist within these dimensions. So…

It is theorized that there are many universes, existing side by side with our own, and even occupying the same space as our own universe in some ethereal manner. Some consider the possibility that these parallel universes exist so close to one another that the gravity is a weak signal leaking out of another universe into ours.

In the 1990s it was suggested that interacting strings in ten dimensions might have an equivalent in weakly-interacting five-dimensional branes.

As expected, physicists were unable to prove this relationship for two important reasons: For one, the Montonen–Olive duality was still unproven, and so the speculation was even more tenuous. Alternatively, there were many technical issues connected to the quantum properties of five-dimensional branes, the first of which was not solved until it was proven that certain physical theories require the existence of objects with both electric and magnetic charge.

Remember that a cornerstone theory of parapsychology purports that ghosts or otherworldly entities use electromagnetic forces to communicate with humans on this mortal coil we call existence, which is why ghost hunters experiment with and measure electromagnetic fields.

Now, to simplify all of this, membrane theory posits that the universe is connected via electromagnetic layers of energy and to each other universe by the same.

Since there is much we do not yet understand about our own universe, let alone any others, it is well within the realm of reason that what we think of as life after death and spirit manifestation is simultaneously our life force, or energy, returning to this collective and beings from other dimensions, or the non-corporeal remnants of humans passing in between what we perceive of existence through thin areas between these membranes.

Think about it.

If a bolt of lightning can pass through objects several feet thick, then why couldn't that same energy pass through when two universes that are thousandths of a millimeter apart bump into each other?

It has been suggested that the Big-Bang that created our universe was the result of two universes touching. If so, then the Big Bang wasn't really the beginning of everything and that time and space all existed before it, and will again–like waves receding and crashing along the shore.

If you were to put all of this into practice it would be similar to (and completely in line with quantum mechanics) randomly appearing in more than one place at once before shrinking and exploding into nothingness only to reappear at another place and time and you're suddenly walking a cow in ancient Egypt.

It may seem like I've lost my mind, but this is exactly what quantum mechanics involves. For a quick video to explain all of this visually, this is a great link: http://youtu.be/JkxieS-6WuA.

There you have it. By removing all of the stigmatic and clichéd ghost terms, mainstream science has proven the possibility of hauntings. What we perceive as a haunting is merely the interaction, however brief, between two dimensions or universes and those individuals inhabiting them.

Brian Josephson, a Noble Prize member of the University of Cambridge, stated in an interview that, "If something unorthodox comes up scientists ignore it...they then say it is nonsense....and then they say it is obvious. I believe in the future, the paranormal might become excepted science if some of the Quantum Physics speculations are found to be correct and then it will become 'obvious'."

As we in the scientific community like to say when such a *eureka!* moment occurs, "That's cool."

The age-old mystics and healers are chuckling, though. Science, it seems, always has to prove what they have known for millennia; but...truth, as they say, is truth. If one takes the long road and another takes the short road, what does it matter so long as both arrive together and enjoy the adventures along the way.

The future, at least for paranormal research, is once again a wondrous stroll through the possible.

Original Publication Order

Dec 2009–Parapsychology Today
Jan 2010–Experimenting with Electronic Voice Phenomena
Feb 2010–Do You See What I See: Spirit Photography
Mar 2010–Spiritual Work and Paranormal Investigations
Apr 2010–Spiritual Work and Paranormal Investigations (2 of 3)
May 2010–Spiritual Work and Paranormal Investigations (3 of 3)
Jun 2010–Debunking Paranormal TV Shows
Jul 2010–Michigan Hauntings
Aug 2010–Investigation Procedures (Part 1 of 5)
Sep 2010–Investigation Procedures (Part 2 of 5)
Oct 2010–Investigation Procedures (Part 3 of 5)
Nov 2010–Investigation Procedures (Part 4 of 5)
Dec 2010–Investigation Procedures (Part 5 of 5)

Jan 2011–An exploration of Near-Death and Out-of-Body Experiences
Feb 2011–Springtime Calls Ghost Hunters Back Outdoors
Mar 2011–The Embodied Mind: Altered States of Consciousness
Apr 2011–Ghost Hunting and Entertainment
May 2011–Crisis Apparitions
Jun 2011–Environmental Factors of Ghost Hunting
Jul 2011–Environmental Factors of Ghost Hunting: The Moon
Aug 2011–Environmental Factors of Ghost Hunting: The Sun
Sep 2011–The Ghostly Side of Michigan State University
Oct 2011–Harvests and Hauntings–Autumn in Michigan
Nov 2011–Science and Psychics–The Tech of Paranormal Research
Dec 2011–The Ghosts beneath the Mistletoe

Jan 2012–The Harsh Truth about Ghost Boxes
Feb 2012–Do We Need Parapsychology?
Mar 2012–"Hey Can You Look at This?!"
Apr 2012–The Ghost of Belle Isle
May 2012–The Curse of Peche Island
Jun 2012–Haunted House for Rent
Jul 2012–Hocus Pocus for Profit
Aug 2012–The Perception of Believing
Sep 2012–Ghost Hunting Doesn't Involve Breaking the Law
Oct 2012–Innovations in Paranormal Tech
Nov 2012–Haunted Hotels

Dec 2012–The Gray Lady

Jan 2013–The Unanswered Question
Feb 2013–What does science have to fear from parapsychology?
Mar 2013–Packing Up and Moving a Haunted House
Apr 2013–Some Debts are Hard to Pay
May 2013–The Minefield between Paranormal Belief and Religion
Jun 2013–Paranormal Healing
Jul 2013–Stump the Ghost Guy
Aug 2013–When Fantasy Meets Reality: The Conjuring
Sep 2013–Ghost Hunting: A Eulogy
Oct 2013–When Ghost Hunters Go shopping
Nov 2013–Who Left the Gate Open? The Idiots Got Out Again.

Jan 2014–Where, Oh Where, Has the Science Gone?
Feb 2014–Paranormal Communication
Mar 2014 - Parapsychology's Database Debacle
Apr 2014–The only thing worse than amateur scientists
May 2014 ---
Jun 2014–The Fine Line Between Believer and Skeptic
July 2014–All Good Things (written May 2014)